Edexcel AS and A Level Modular Mathematics

Core Mathematics 4

Greg Attwood
Alistair Macpherson
Bronwen Moran
Joe Petran
Keith Pledger
Geoff Staley
Dave Wilkins

Published by Pearson Education Limited, a company incorporated in England and Wales, having its registered office at Edinburgh Gate, Harlow, Essex, CM20 2JE. Registered company number: 872828

Edexcel is a registered trademark of Edexcel Limited

First published 2009

14
10 9 8 7

British Library Cataloguing in Publication Data is available from the British Library on request.

ISBN 978 0 435519 29 2

Edited by Susan Gardner
Typeset by Tech-Set Ltd
Illustrated by Tech-Set Ltd
Cover design by Christopher Howson
Picture research by Chrissie Martin
Cover photo/illustration © Edexcel
Printed in Malaysia, CTP-PJB

Acknowledgements
Every effort has been made to contact copyright holders of material reproduced in this book.
Any omissions will be rectified in subsequent printings if notice is given to the publishers.

About this book

Revise for Core 4 covers the key topics that are tested in the Core 4 examination paper. You can use this book to help you revise at the end of your course, or you can use it throughout your course alongside the course textbook, *Edexcel AS and A Level Modular Mathematics Core 4*, which provides complete coverage of the specification.

Helping you prepare for your examination

To help you prepare, each topic offers you:

- **What you should know** – a summary of the mathematical ideas you need to know and be able to use.
- **Test Yourself questions** – help you see where you need extra revision and practice. If you do need extra help, they show you where to look in the *Edexcel AS and A Level Modular Mathematics Core 4* textbook and which example to refer to in this book.
- **Worked examples and examination questions** – help you understand and remember important methods, and show you how to set out your answers clearly.
- **Revision exercises** – help you practise using these important methods to solve problems. Examination-level questions are included so that you can be sure that you are reaching the right standard, and answers are given at the back of the book so that you can assess your progress.

Examination practice and advice on revising

Examination style paper – this paper at the end of the book provides a set of questions of examination standard. It gives you an opportunity to practise taking a complete examination before you meet the real thing. The answers are given at the back of the book.

How to revise – for advice on revising before the examination, read the How to revise section on the next page.

How to revise using this book

Making the best use of your revision time

The topics in this book have been arranged in a logical sequence so that you can work your way through them from beginning to end. However, **how** you work on them depends on how much time there is between now and your examination.

If you have plenty of time before the examination then you can **work through each topic in turn**, covering the what you should know section and worked examples before doing the revision exercises.

If you are short of time then you can **work through the Test Yourself sections** first, to help you see which topics you need to do further work on.

However much time you have to revise, make sure you break your revision into short blocks of about 40 minutes, separated by five- or ten-minute breaks. Nobody can study effectively for hours without a break.

Using the Test Yourself sections

Each Test Yourself section provides a set of key questions. Try each question.

- If you can do it and get the correct answer, then move on to the next topic. Come back to this topic later to consolidate your knowledge and understanding by working through the what you should know section, worked examples and revision exercises.
- If you cannot do the question, or get an incorrect answer or part answer, then work through the what you should know section, worked examples and revision exercises before trying the Test Yourself questions again. If you need more help, the cross-references beside each Test Yourself question show you where to find relevant information in the *Edexcel AS and A Level Modular Mathematics Core 4* textbook and which example in *Revise for C4* to refer to.

Reviewing the what you should know sections

Most of the things that you should know are straightforward ideas that you can learn: try to understand each one. Imagine explaining each idea to a friend in your own words, and say it out loud as you do so. This is a better way of making the ideas stick than just reading them silently from the page.

As you work through the book, remember to go back over the things that you should know from earlier topics at least once a week. This will help you to remember them in the examination.

Partial fractions 1

What you should know

1 An algebraic fraction can be written as a sum of two or more simpler fractions. This technique is called splitting into partial fractions.

2 An expression with two linear terms in the denominator such as $\frac{11}{(x-3)(x+2)}$ can be split by converting into the form $\frac{A}{(x-3)} + \frac{B}{(x+2)}$.

3 An expression with three or more linear terms such as $\frac{4}{(x+1)(x-3)(x+4)}$ can be split by converting into the form $\frac{A}{(x+1)} + \frac{B}{(x-3)} + \frac{C}{(x+4)}$ and so on if there are more terms.

4 An expression with repeated terms in the denominator such as $\frac{6x^2 - 29x - 29}{(x+1)(x-3)^2}$ can be split by converting into the form $\frac{A}{(x+1)} + \frac{B}{(x-3)} + \frac{C}{(x-3)^2}$.

5 An improper fraction is one where the index of the numerator is equal to or higher than the index of the denominator. An improper fraction must be divided first to obtain a number and a proper fraction before you can express it in partial fractions.

- For example,

$$\frac{x^2 + 3x + 4}{x^2 + 3x + 2} = 1 + \frac{2}{x^2 + 3x + 2} = 1 + \frac{A}{(x+1)} + \frac{B}{(x+2)}.$$

Test yourself

1 Express $\dfrac{x-31}{(x+5)(x-4)}$ in partial fraction form.

2 Express $\dfrac{4x^2-28}{(x+2)^2(x-4)}$ in the form

$\dfrac{A}{(x+2)^2}+\dfrac{B}{(x+2)}+\dfrac{C}{(x-4)}$.

3 Show that $\dfrac{x^3-4x^2+5x+2}{x^2-2x-3}$ can be put in the form

$Ax+B+\dfrac{C}{(x-3)}+\dfrac{D}{(x+1)}$ where A, B, C and D are constants to be determined.

4 Given that $y=\dfrac{9x^2-2x-3}{x^3-x}$

(a) show that $y \equiv \dfrac{A}{x}+\dfrac{B}{(x+1)}+\dfrac{C}{(x-1)}$ where A, B and C are constants to be determined.

(b) Hence or otherwise, find $\dfrac{\mathrm{d}y}{\mathrm{d}x}$ and show that the gradient of the curve at $x=3$ is $-\frac{13}{12}$.

What to review

If your answer is incorrect

1. *Review Edexcel Book C4 pages 3–4*
   Revise for C4 page 3 Example 1

2. *Review Edexcel Book C4 page 6*
   Revise for C4 page 4 Example 2

3. *Review Edexcel Book C4 page 7*
   Revise for C4 page 5 Example 3

4. (a) *Review Edexcel Book C4 page 5*
   Revise for C4 page 6 Worked examination style question 1a

   (b) *Review Edexcel Book C3 pages 133–134*

Example 1

Express $\dfrac{5x-10}{(x-4)(x+1)}$ in partial fraction form.

Let $\dfrac{5x-10}{(x-4)(x+1)} \equiv \dfrac{A}{(x-4)} + \dfrac{B}{(x+1)}$

Using **2**: there are two linear terms in the denominator.

$$\equiv \frac{A(x+1)+B(x-4)}{(x-4)(x+1)}$$

Add the fractions.

Hence $5x-10 \equiv A(x+1)+B(x-4)$

Set the numerators equal to each other.

Let $x=4$ $\quad 5\times 4-10 = A(4+1)+B\times 0$

$$10 = 5A$$

$$\underline{A = 2}$$

To find A substitute $x = 4$.

Let $x=-1$ $\quad 5\times -1-10 = A\times 0 + B(-1-4)$

$$-15 = -5B$$

$$\underline{B = 3}$$

To find B substitute $x = -1$.

$$\therefore \quad \frac{5x-10}{(x-4)(x+1)} \equiv \frac{2}{(x-4)} + \frac{3}{(x+1)}$$

Example 2

Split $\dfrac{-7x^2 - 4x}{(x + 1)^2(2x + 1)}$ into partial fractions.

Let $\dfrac{-7x^2 - 4x}{(x + 1)^2(2x + 1)}$

$$\equiv \frac{A}{(x+1)} + \frac{B}{(x+1)^2} + \frac{C}{(2x+1)}$$

Using **4**: the $(x + 1)$ term is repeated.

$$\equiv \frac{A(x+1)(2x+1) + B(2x+1) + C(x+1)^2}{(x+1)^2(2x+1)}$$

Add the fractions.

$$-7x^2 - 4x \equiv A(x+1)(2x+1) + B(2x+1) + C(x+1)^2$$

Set the numerators equal to each other.

Let $x = -1$ $\quad -7 + 4 = A \times 0 + B \times -1 + C \times 0$

$-3 = -1B$

$\underline{B = 3}$

To find B substitute $x = -1$.

Let $x = -\frac{1}{2}$ $\quad -\frac{7}{4} + 2 = A \times 0 + B \times 0 + C \times \frac{1}{4}$

$\frac{1}{4} = \frac{1}{4}C$

$\underline{C = 1}$

To find C substitute $x = -\frac{1}{2}$.

Comparing terms in x^2 $\quad -7 = A \times 2 + C \times 1$

To find A equate terms in x^2.

$-7 = 2A + 1$

Substitute $C = 1$.

$2A = -8$

$\underline{A = -4}$

$$\therefore \frac{-7x^2 - 4x}{(x+1)^2(2x+1)} \equiv -\frac{4}{(x+1)} + \frac{3}{(x+1)^2} + \frac{1}{(2x+1)}$$

1

Example 3

Express $\frac{x^3 - x^2 - 3x - 4}{x^2 - x - 2}$ in partial fraction form.

As the numerator has index = 3 and denominator has index = 2, this is an improper fraction.

$$\begin{array}{r} x \\ x^2 - x - 2\overline{)x^3 - x^2 - 3x - 4} \\ \underline{x^3 - x^2 - 2x} \\ -x - 4 \end{array}$$

Using **5**: divide the denominator into the numerator.

This is the remainder.

Therefore $\frac{x^3 - x^2 - 3x - 4}{x^2 - x - 2} \equiv x + \frac{-x - 4}{x^2 - x - 2}$

$$\equiv x + \frac{-x - 4}{(x - 2)(x + 1)}$$

Factorise the denominator.

Let $\frac{-x - 4}{(x - 2)(x + 1)} \equiv \frac{A}{(x - 2)} + \frac{B}{(x + 1)}$

Using **2**: express the remainder in partial fraction form.

$$\frac{-x - 4}{(x - 2)(x + 1)} \equiv \frac{A(x + 1) + B(x - 2)}{(x - 2)(x + 1)}$$

Add the fractions.

$$-x - 4 \equiv A(x + 1) + B(x - 2)$$

Set the numerators equal to each other.

Let $x = 2$ $\quad -2 - 4 = A \times 3 + B \times 0$

$$3A = -6$$

$$\underline{A = -2}$$

To find A substitute $x = 2$.

Let $x = -1$ $\quad 1 - 4 = A \times 0 + B \times -3$

$$-3B = -3$$

$$\underline{B = +1}$$

To find B substitute $x = -1$.

$$\therefore \quad \frac{x^3 - x^2 - 3x - 4}{x^2 - x - 2} \equiv x - \frac{2}{(x - 2)} + \frac{1}{(x + 1)}$$

Worked examination style question 1

Given that $\dfrac{-8x-2}{(x+1)x(x-2)} \equiv \dfrac{A}{(x+1)} + \dfrac{B}{x} + \dfrac{C}{(x-2)}$

(a) find the values of the constants A, B and C.

(b) Hence or otherwise show that $\displaystyle\int_3^4 \frac{-8x-2}{(x+1)x(x-2)}\,\mathrm{d}x = \ln\left(\frac{25}{96}\right)$.

(a) Let $\dfrac{-8x-2}{(x+1)x(x-2)}$

$$\equiv \frac{A}{(x+1)} + \frac{B}{x} + \frac{C}{(x-2)}$$

Using **3**: there are three linear terms on the denominator.

$$\equiv \frac{Ax(x-2) + B(x+1)(x-2) + Cx(x+1)}{(x+1)x(x-2)}$$

Add the fractions.

$$-8x-2 \equiv Ax(x-2) + B(x+1)(x-2) + Cx(x+1)$$

Set the numerators equal to each other.

Let $x = 0$ $\quad -2 = A \times 0 + B \times 1 \times -2 + C \times 0$

$-2 = -2B$

$\underline{B = 1}$

Substitute $x = 0$ to find B.

Let $x = 2$ $\quad -18 = A \times 0 + B \times 0 + C \times 2 \times 3$

$-18 = 6C$

$\underline{C = -3}$

Substitute $x = 2$ to find C.

Let $x = -1$ $\quad 6 = A \times -1 \times -3 + B \times 0 + C \times 0$

$6 = 3A$

$\underline{A = 2}$

Substitute $x = -1$ to find A.

$$\therefore \quad \frac{-8x-2}{(x+1)x(x-2)} \equiv \frac{2}{x+1} + \frac{1}{x} - \frac{3}{(x-2)}$$

1

(b) $\int_3^4 \frac{-8x-2}{(x+1)x(x-2)}\,dx$

$\equiv \int_3^4 \frac{2}{(x+1)} + \frac{1}{x} - \frac{3}{(x-2)}\,dx$

$= \left[2\ln|x+1| + \ln|x| - 3\ln|x-2|\right]_3^4$

$= (2\ln 5 + \ln 4 - 3\ln 2) - (2\ln 4 + \ln 3 - 3\ln 1)$

$= 2\ln 5 + \ln 4 - 3\ln 2 - 2\ln 4 - \ln 3$

$= \ln 5^2 + \ln 4 - \ln 2^3 - \ln 4^2 - \ln 3$

$= \ln 5^2 + \ln 2^2 - (\ln 2^3 + \ln 2^4 + \ln 3)$

$= \ln\left(\frac{5^2 \times 2^2}{2^3 \times 2^4 \times 3}\right)$

$= \ln\left(\frac{5^2}{2^5 \times 3}\right)$

$= \ln\left(\frac{25}{96}\right)$

Write the integral in partial fraction form.

Integrate each term separately: $\int \frac{1}{x}\,dx = \ln|x|$.

$\ln 1 = 0$.

$n \ln a = \ln a^n$.

Use $\ln a + \ln b = \ln(ab)$.

$\ln a - \ln b = \ln\left(\frac{a}{b}\right)$.

Cancel $\frac{4}{2^3} = \frac{1}{2}$.

Worked examination style question 2

The function f(x) is defined by

$$f(x) = \frac{3 + 7x}{(1 + x)(1 + 2x)}$$

(a) express f(x) in the form $\frac{A}{(1 + x)} + \frac{B}{(1 + 2x)}$ where A and B are constants which should be found.

(b) Hence or otherwise find in ascending powers of x, the binomial expansion of f(x) up to and including the terms in x^3.

(a) Let $\frac{3 + 7x}{(1 + x)(1 + 2x)} \equiv \frac{A}{(1 + x)} + \frac{B}{(1 + 2x)}$ — Using **2**: the terms $(1 + x)$ and $(1 + 2x)$ are linear.

$$\equiv \frac{A(1 + 2x) + B(1 + x)}{(1 + x)(1 + 2x)}$$ — Add the fractions.

$$3 + 7x \equiv A(1 + 2x) + B(1 + x)$$ — Set the numerators equal to each other.

Substitute $x = -1$ $\quad 3 - 7 = A \times (-1) + B \times 0$ — To find A substitute $x = -1$.

$$A = 4$$

Substitute $x = -\frac{1}{2}$ $\quad 3 - 3\frac{1}{2} = A \times 0 + B \times \frac{1}{2}$ — To find B substitute $x = -\frac{1}{2}$.

$$B = -1$$

Hence $\quad f(x) = \frac{4}{(1 + x)} - \frac{1}{(1 + 2x)}$

(b) $f(x) = 4(1 + x)^{-1} - (1 + 2x)^{-1}$ — Write in index form.

$= 4(1 - x + x^2 - x^3 + \ldots)$
$\quad - (1 - 2x + 4x^2 - 8x^3 + \ldots)$ — Use Chapter 3, **1**.

$= 4 - 4x + 4x^2 - 4x^3$
$\quad - 1 + 2x - 4x^2 + 8x^3 + \ldots$ — Expand the brackets.

$= 3 - 2x + 4x^3 + \ldots$ — Collect 'like terms'.

Revision exercise 1

1 Express $\dfrac{-x-11}{(2x+1)(x-3)}$ in partial fraction form.

2 Express $\dfrac{3x-2}{(x-2)^2}$ in the form $\dfrac{A}{(x-2)}+\dfrac{B}{(x-2)^2}$ where A and B are constants to be found.

3 Express the improper fraction $\dfrac{x^3+5x^2+x-24}{x^2+2x-8}$ in partial fraction form.

4 Given that $\dfrac{3x-11}{(x+3)(x-2)} \equiv \dfrac{A}{(x+3)}+\dfrac{B}{(x-2)}$

(a) find the value of the constants A and B.

(b) Hence or otherwise find the gradient of the curve

$y = \dfrac{3x-11}{(x+3)(x-2)}$ at the point where $x = -1$.

5 Given that $\dfrac{6x^2+4x-1}{(x+2)(2x-1)(x-1)} \equiv \dfrac{A}{(x+2)}+\dfrac{B}{(2x-1)}+\dfrac{C}{(x-1)}$

(a) find the value of the constants A, B and C.

(b) Hence or otherwise prove that

$$\int_2^3 \frac{6x^2+4x-1}{(x+2)(2x-1)(x-1)}\,dx = \ln\left(\frac{2}{3}\right).$$

6 Given that $\dfrac{5x^2-17x}{(x-2)^2(2x+3)} \equiv \dfrac{A}{(x-2)}+\dfrac{B}{(x-2)^2}+\dfrac{C}{(2x+3)}$

(a) find the value of the constants A, B and C.

(b) Hence or otherwise prove that the gradient of the curve

$y = \dfrac{5x^2-17x}{(x-2)^2(2x+3)}$ at the point $x = \frac{3}{2}$ is $-\frac{217}{6}$.

7 Given that $\dfrac{x+14}{(2+x)(1-x)} \equiv \dfrac{A}{(2+x)} + \dfrac{B}{(1-x)}$

(a) find the value of the constants A and B.

(b) Hence or otherwise find the first three terms in the binomial expansion of $\dfrac{x+14}{(2+x)(1-x)}$.

(c) State the values of x for which this expansion is valid.

8 The function f(x) is defined by

$$f(x) = \frac{6x^2 - 27x + 29}{(x-3)(2-x)^2} \quad \{x \in \mathbb{R}, x > 3\}$$

(a) Show that f(x) can be written in the form

$$\frac{A}{(x-3)} + \frac{B}{(2-x)} + \frac{C}{(2-x)^2}.$$

(b) Hence or otherwise show that $f(x) > 0$ for all values of the domain.

9 Given that $\dfrac{11x-1}{(1-x)^2(2+3x)} \equiv \dfrac{A}{(1-x)^2} + \dfrac{B}{(1-x)} + \dfrac{C}{(2+3x)}$

(a) find the values of A, B and C.

(b) Find the exact value of $\displaystyle\int_0^{\frac{1}{2}} \frac{11x-1}{(1-x)^2(2+3x)}\,dx$, giving your answer in the form $k + \ln a$, where k is an integer and a is a simplified fraction.

E

Coordinate geometry in the (x, y) plane

2

What you should know

1 To find the Cartesian equation of a curve given parametrically you eliminate the parameter between the parametric equations.

2 To find the area under a curve given parametrically you use $\int y \frac{dx}{dt}\, dt$.

Test yourself	What to review
	If your answer is incorrect
1 Find the Cartesian equation of the curve given by the parametric equations $x = t^{\frac{3}{2}}$, $y = 4t^3$.	*Review Edexcel Book C4 page 12* *Revise for C4 page 12 Example 1*
2 The curve with parametric equations $x = 4t$, $y = \frac{4}{t}$, $t \neq 0$ meets the curve $2y^2 + x = 0$ at P. Find the coordinates of the point P.	*Review Edexcel Book C4 page 14* *Revise for C4 page 15 Example 3*
3 A circle has parametric equations $x = \cos\theta - 5$, $y = \sin\theta + 3$, $0 \leqslant \theta \leqslant 2\pi$. Find the Cartesian equation of the circle.	*Review Edexcel Book C4 page 16* *Revise for C4 page 13 Example 2*
4 A curve has parametric equations $x = t^2 - 1$, $y = \frac{1}{2}(t - t^3)$. **(a)** Draw a graph of the curve for $-2 \leqslant t \leqslant 2$. **(b)** Find the area of the finite region enclosed by the loop of the curve.	*Review Edexcel Book C4 pages 11 and 18* *Revise for C4 page 16 Example 4*

Example 1

A curve is given parametrically by the equations

$x = \frac{1}{t+1}$, $y = \frac{1}{t-2}$, $t \neq -1$, $t \neq 2$.

(a) Find an equation for t in terms of x.

(b) Show that the Cartesian equation of the curve is $y = \frac{x}{1-3x}$.

(a) $$x = \frac{1}{t+1}$$

So $$x \times (t+1) = \frac{1}{t+1} \times (t+1)$$

Rearrange $x = \frac{1}{t+1}$ for t.
Multiply each side by $(t+1)$.

$$x(t+1) = 1$$

$$\frac{x(t+1)}{x} = \frac{1}{x}$$

Divide each side by x.

$$t + 1 = \frac{1}{x}$$

So $$t = \frac{1}{x} - 1$$

(b) $$y = \frac{1}{\left(\frac{1}{x} - 1\right) - 2}$$

Substitute $t = \frac{1}{x} - 1$ into $y = \frac{1}{t-2}$ and simplify.

$$= \frac{1}{\frac{1}{x} - 3}$$

$$= \frac{1}{\frac{1}{x} - \frac{3x}{x}}$$

Make $\frac{1}{x} - 3$ into a fraction.

$$= \frac{1}{\left(\frac{1-3x}{x}\right)}$$

The Cartesian equation of the curve is $y = \frac{x}{1-3x}$

Remember $\frac{1}{\left(\frac{a}{b}\right)} = \frac{b}{a}$.

Example 2

The diagram shows a sketch of the ellipse with parametric equations

$x = 2 - 4\cos\theta,\ y = 3\sin\theta + 4,\ 0 \leqslant \theta < 2\pi.$

The ellipse meets the y-axis at A and B.

(a) Find the exact coordinates of the points A and B.

(b) Find a Cartesian equation of the ellipse.

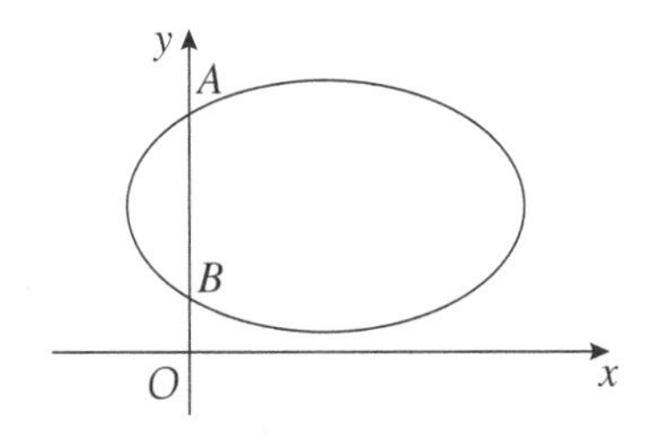

2

(a) $2 - 4\cos\theta = 0$

$4\cos\theta = 2$

$\cos\theta = \frac{1}{2}$

$\theta = \cos^{-1}\left(\frac{1}{2}\right)$

So $\theta = \frac{\pi}{3},\quad \theta = \frac{5\pi}{3}$

For $\theta = \frac{\pi}{3}$

$y = 3\sin\left(\frac{\pi}{3}\right) + 4$

$= 3\left(\frac{\sqrt{3}}{2}\right) + 4$

$= \frac{3\sqrt{3}}{2} + 4$

For $\theta = \frac{5\pi}{3}$

$y = 3\sin\left(\frac{5\pi}{3}\right) + 4$

$= 3\left(-\frac{\sqrt{3}}{2}\right) + 4$

$= 4 - \frac{3\sqrt{3}}{2}$

The ellipse meets the y-axis when $x = 0$, so substitute $x = 0$ into $x = 2 - 4\cos\theta$ and solve for θ.

Remember θ is in radians.

Find the coordinates of A and B, by substituting $\theta = \frac{\pi}{3}$ and $\theta = \frac{5\pi}{3}$ into $y = 3\sin\theta + 4$.

Remember $\sin\frac{\pi}{3} = \frac{\sqrt{3}}{2}$ and $\sin\frac{5\pi}{3} = -\frac{\sqrt{3}}{2}$.

The coordinates are $A\left(0, \frac{3\sqrt{3}}{2} + 4\right)$

and $B\left(0, 4 - \frac{3\sqrt{3}}{2}\right)$

(b) $x + 4\cos\theta = 2$

$4\cos\theta = 2 - x$

$\cos\theta = \frac{2 - x}{4}$

$y = 3\sin\theta + 4$

$3\sin\theta = y - 4$

$\sin\theta = \frac{y - 4}{3}$

As $\cos^2\theta + \sin^2\theta = 1$

the Cartesian equation of the ellipse is

$$\left(\frac{2 - x}{4}\right)^2 + \left(\frac{y - 4}{3}\right)^2 = 1$$

Using **1**: eliminate θ between the equations $x = 2 - 4\cos\theta$ and $y = 3\sin\theta + 4$.

Rearrange $x + 4\cos\theta = 2$ for $\cos\theta$.

Rearrange $y = 3\sin\theta + 4$ for $\sin\theta$.

Square $\cos\theta$ and $\sin\theta$ so that

$\cos^2\theta = \left(\frac{2 - x}{4}\right)^2$

$\sin^2\theta = \left(\frac{y - 4}{3}\right)^2$.

Example 3

The line $x + y = 9$ meets the curve with parametric equations $x = t^2$, $y = (t + 3)(t - 2)$ at P and Q.

Find the coordinates of the points P and Q.

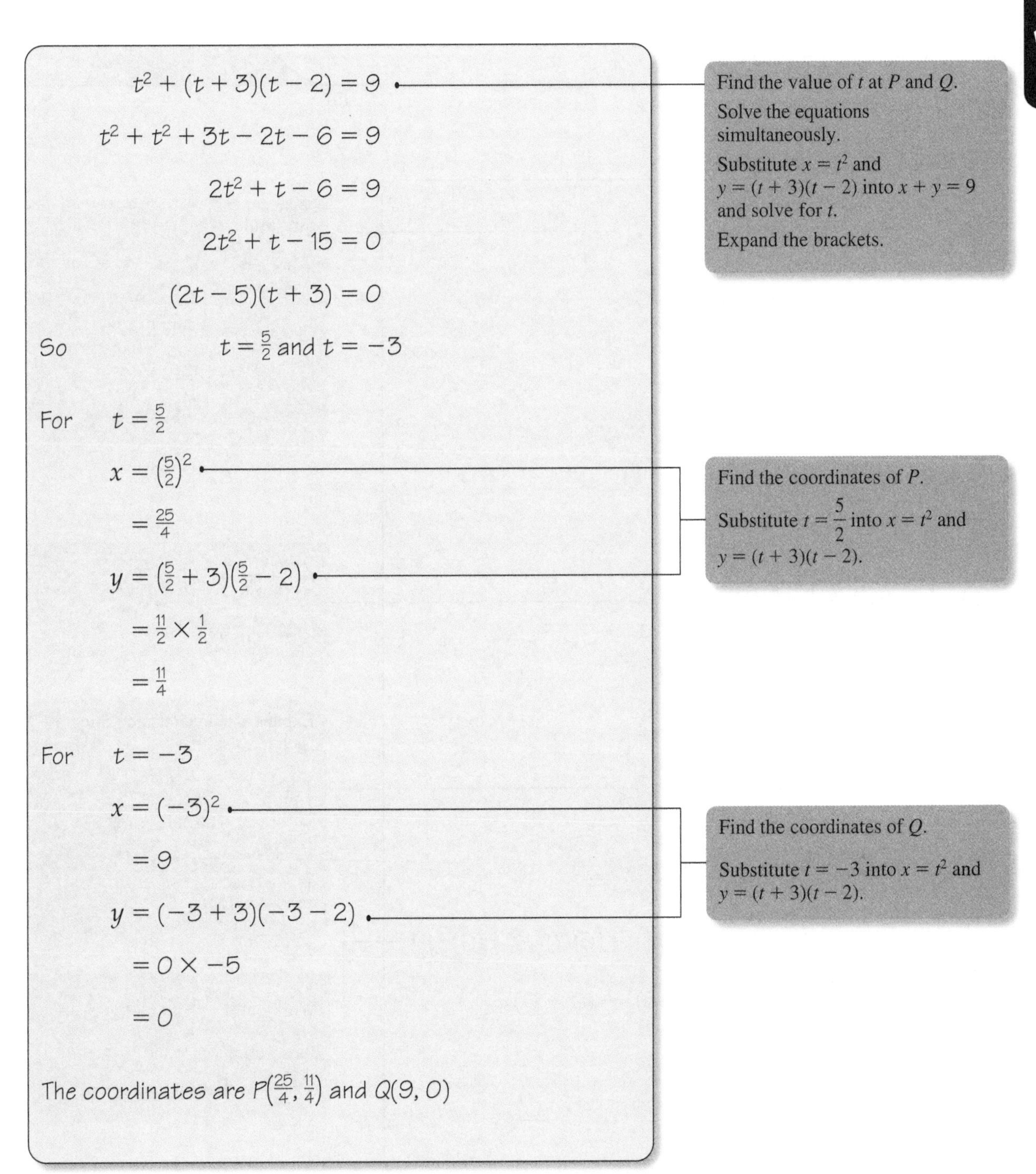

$$t^2 + (t + 3)(t - 2) = 9$$
$$t^2 + t^2 + 3t - 2t - 6 = 9$$
$$2t^2 + t - 6 = 9$$
$$2t^2 + t - 15 = 0$$
$$(2t - 5)(t + 3) = 0$$

So $t = \frac{5}{2}$ and $t = -3$

Find the value of t at P and Q.

Solve the equations simultaneously.

Substitute $x = t^2$ and $y = (t + 3)(t - 2)$ into $x + y = 9$ and solve for t.

Expand the brackets.

For $t = \frac{5}{2}$

$$x = \left(\tfrac{5}{2}\right)^2 = \tfrac{25}{4}$$
$$y = \left(\tfrac{5}{2} + 3\right)\left(\tfrac{5}{2} - 2\right) = \tfrac{11}{2} \times \tfrac{1}{2} = \tfrac{11}{4}$$

Find the coordinates of P.

Substitute $t = \dfrac{5}{2}$ into $x = t^2$ and $y = (t + 3)(t - 2)$.

For $t = -3$

$$x = (-3)^2 = 9$$
$$y = (-3 + 3)(-3 - 2) = 0 \times -5 = 0$$

Find the coordinates of Q.

Substitute $t = -3$ into $x = t^2$ and $y = (t + 3)(t - 2)$.

The coordinates are $P\left(\frac{25}{4}, \frac{11}{4}\right)$ and $Q(9, 0)$

Example 4

The diagram shows a sketch of the curve with parametric equations

$x = 2t^{\frac{1}{2}}, y = t(9 - t), 0 \leqslant t \leqslant 9.$

The shaded region R is bounded by the curve and the x-axis.

(a) Show that $y\dfrac{dx}{dt} = t^{\frac{1}{2}}(9 - t)$.

(b) Find the area of R.

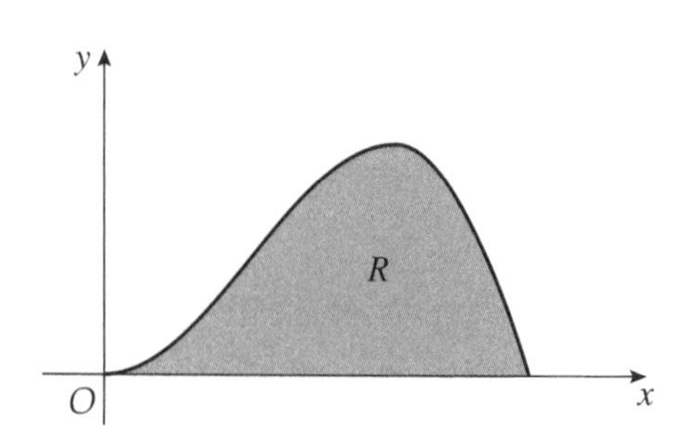

(a)

$$y\frac{dx}{dt} = t(9-t)\frac{d}{dt}(2t^{\frac{1}{2}})$$

$$= t(9-t) \times t^{-\frac{1}{2}}$$

$$= t^{1-\frac{1}{2}}(9-t)$$

$$= t^{\frac{1}{2}}(9-t)$$

So $y\dfrac{dx}{dt} = t^{\frac{1}{2}}(9-t)$

Substitute $y = t(9 - t)$ and $x = 2t^{\frac{1}{2}}$ into $y\dfrac{dx}{dt}$.

$$\frac{dx}{dt} = \frac{d}{dt}(2t^{\frac{1}{2}}) = \tfrac{1}{2} \times 2t^{\frac{1}{2}-1} = t^{-\frac{1}{2}}$$

(b)

$$\int y\frac{dx}{dt}\,dt = \int_0^9 t^{\frac{1}{2}}(9-t)\,dt$$

Using **2**: $R = \int y\dfrac{dx}{dt}\,dt$. Here $y\dfrac{dx}{dt} = t^{\frac{1}{2}}(9 - t)$.

$$= \int_0^9 9t^{\frac{1}{2}} - t^{\frac{3}{2}}\,dt$$

$$= \left[\frac{9}{\left(\frac{3}{2}\right)}t^{\frac{3}{2}} - \frac{t^{\frac{5}{2}}}{\left(\frac{5}{2}\right)}\right]_0^9$$

Expand the brackets and integrate each term so that

$$\int 9t^{\frac{1}{2}}\,dt = \frac{9t^{\frac{1}{2}+1}}{\left(\frac{1}{2}+1\right)} = \frac{9t^{\frac{3}{2}}}{\left(\frac{3}{2}\right)}$$

$$\int t^{\frac{3}{2}}\,dt = \frac{t^{\frac{3}{2}+1}}{\left(\frac{3}{2}+1\right)} = \frac{t^{\frac{5}{2}}}{\left(\frac{5}{2}\right)}.$$

$$= \left[6t^{\frac{3}{2}} - \tfrac{2}{5}t^{\frac{5}{2}}\right]_0^9$$

$$= \left(6(9)^{\frac{3}{2}} - \tfrac{2}{5}(9)^{\frac{5}{2}}\right) - \left(6(0)^{\frac{3}{2}} - \tfrac{2}{5}(0)^{\frac{5}{2}}\right)$$

$$= \left(6(27) - \tfrac{2}{5}(243)\right) - (0 - 0)$$

$$= 64\tfrac{4}{5}$$

Work out $\left[6t^{\frac{3}{2}} - \tfrac{2}{5}t^{\frac{5}{2}}\right]_0^9$ by substituting $t = 9$ and $t = 0$ into $6t^{\frac{3}{2}} - \tfrac{2}{5}t^{\frac{5}{2}}$ and subtracting.

The area of R is $64\tfrac{4}{5}$

Example 5

The diagram shows a sketch of the ellipse with parametric equations

$x = 4 \sin t,\ y = 3 \cos t,\ 0 \leqslant t \leqslant 2\pi.$

The shaded region R is bounded by the ellipse, $0 \leqslant t \leqslant \frac{\pi}{2}$, and the x-axis.

(a) Find the area of R.

(b) Hence, write down the total area enclosed by the ellipse.

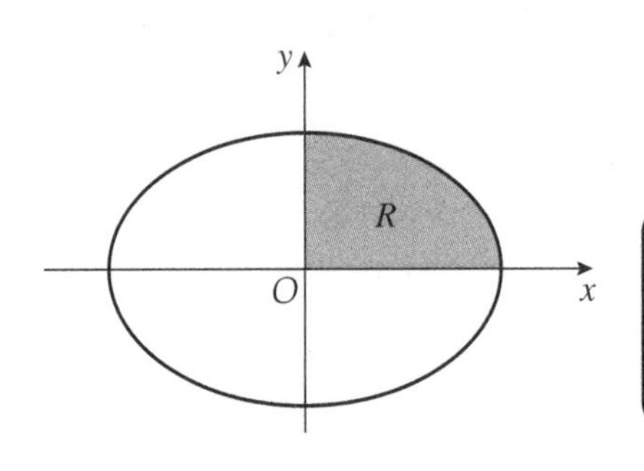

(a)
$$y\frac{dx}{dt} = 3\cos t \frac{d}{dt}(4\sin t)$$
$$= 3\cos t \times 4\cos t$$
$$= 12\cos^2 t$$

Using **2**: $R = \int y\frac{dx}{dt}\,dt$.
Substitute $y = 3\cos t$ into $y\frac{dx}{dt}$ and work out $\frac{dx}{dt}$.

Therefore

$$\int y\frac{dx}{dt}\,dt = \int_0^{\frac{\pi}{2}} 12\cos^2 t\,dt$$
$$= \int_0^{\frac{\pi}{2}} 12 \times \tfrac{1}{2}(\cos 2t + 1)\,dt$$

Rearrange $\cos 2t = 2\cos^2 t - 1$ for $\cos^2 t$.
$2\cos^2 t = \cos 2t + 1$
$\cos^2 t = \frac{1}{2}(\cos 2t + 1)$.

$$= \int_0^{\frac{\pi}{2}} 6(\cos 2t + 1)\,dt$$
$$= \int_0^{\frac{\pi}{2}} 6\cos 2t + 6\,dt$$

Integrate each term so that
$\int \cos 2t\,dt = \frac{1}{2}\sin 2t$
$\int 6\,dt = 6t$.

$$= \left[6 \times \tfrac{1}{2}\sin 2t + 6t\right]_0^{\frac{\pi}{2}}$$
$$= \left[3\sin 2t + 6t\right]_0^{\frac{\pi}{2}}$$
$$= \left(3\sin\left(2 \times \frac{\pi}{2}\right) + 6 \times \frac{\pi}{2}\right) - (3\sin(2\times 0) + 6 \times 0)$$

Work out $\left[3\sin 2t + 6t\right]_0^{\frac{\pi}{2}}$ by substituting $t = \frac{\pi}{2}$ and $t = 0$ into $3\sin 2t + 6t$ and subtracting.
Remember $\sin \pi = \sin 0 = 0$.

$$= (3\sin(\pi) + 3\pi) - (3\sin(0) + 0)$$
$$= (3 \times 0 + 3\pi) - (3 \times 0 + 0)$$
$$= 3\pi$$

The area of $R = 3\pi$

(b) The area enclosed by the ellipse $= 4 \times 3\pi$
$= 12\pi$

By symmetry there are 4 equal areas.
Each area $= R$
So total area $= 4 \times R$.

Revision exercise 2

1 A curve has parametric equations $x = at^2$, $y = a(8t^3 - 1)$, where a is a constant. The curve passes through the point (4, 0).

(a) Find the value of a.

(b) Find a Cartesian equation of the curve.

2 The line $x + 3y + 2 = 0$ meets the curve with parametric equations $x = 4t^2 - 9$, $y = 2t + 1$ at A and B.

(a) Find the coordinates of the points A and B.

(b) Express t in terms of x.

(c) Hence find the Cartesian equation of the curve in the form $(y - 1)^2 = x + p$, where p is a constant to be found.

3 The curve C has parametric equations $x = 4t - 3$, $y = 16t(t - 1)$.

(a) Find the coordinates of the point where C meets the y-axis.

(b) Find the coordinates of the point where C meets the line $y = 2x + 6$.

(c) Show that the Cartesian equation of C can be written in the form $y = (x + a)^2 + b$, where the values of a and b are to be found.

4 A curve has parametric equations $x = p(4t - 3)$, $y = p(27 + 8t^3)$, where p is a constant. The curve meets the x-axis at $(-4, 0)$ and the y-axis at A.

(a) Find the value of p.

(b) Find the coordinates of A.

5 A circle has parametric equations $x = 13(\cos\theta + 1)$, $y = 13\sin\theta + 5$.

(a) Find a Cartesian equation of the circle.

(b) Draw a sketch of the circle.

(c) Find the coordinates of the points of intersection of the circle with the x-axis.

6 The diagram shows a sketch of the curve with parametric equations $x = \sin 2\theta$, $y = 2\cos\theta$, $0 \leqslant \theta \leqslant 2\pi$.

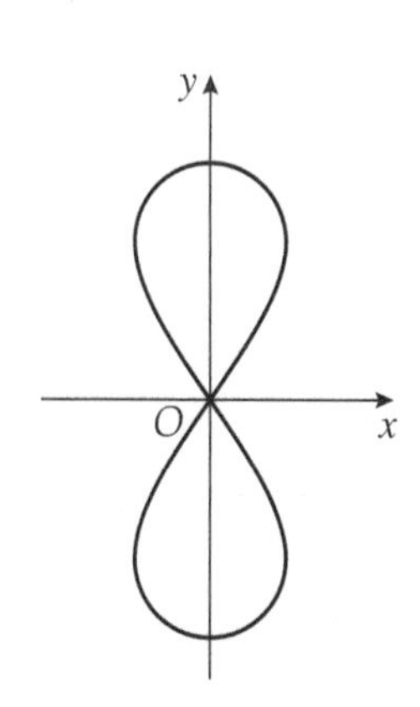

(a) Copy the diagram and label the points having parameters $\theta = 0$, $\theta = \dfrac{\pi}{4}$, $\theta = \pi$.

(b) Find the Cartesian equation of the curve in the form $x^2 = \frac{1}{4}(a^2 - y^2)y^2$, where a is a constant to be found.

7 The diagram shows a sketch of the curve with parametric equations $x = 2\cos\theta$, $y = 3\cos 3\theta$, $0 \leqslant \theta \leqslant 2\pi$.
The curve meets the x-axis at A, O and B.
Find the exact coordinates of the points A and B.

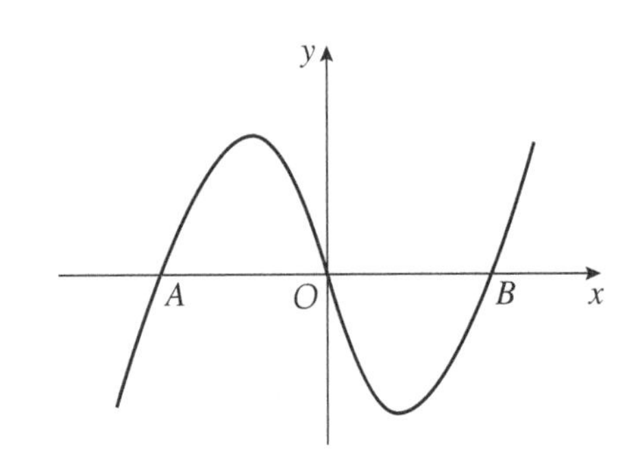

8 The curve with parametric equations $x = 2t^3$, $y = 4t^2 - 9$, meets the line $2x - 3y - 11 = 0$ at A and B.

(a) Show that $t^3 - 3t^2 + 4 = 0$ at the points A and B.

(b) Show that $(t - 2)$ is a factor of $t^3 - 3t^2 + 4$ and hence factorise the expression completely.

(c) Find the coordinates of the points A and B.

9 A line is given parametrically by the equation

$$x = \frac{2t}{t+4},\ y = \frac{1}{t+4},\ t \neq -4.$$

(a) Express t in terms of y.

(b) Find the equation of the line in the form $ax + by + c = 0$, where the values of a, b and c are to be found.

10 The diagram shows a sketch of the curve with parametric equations $x = t^2 - 3$, $y = t^3$, $0 \leqslant t \leqslant 2$.
The curve meets the coordinate axes at P and Q. The region R is bounded by the curve and the coordinate axes.

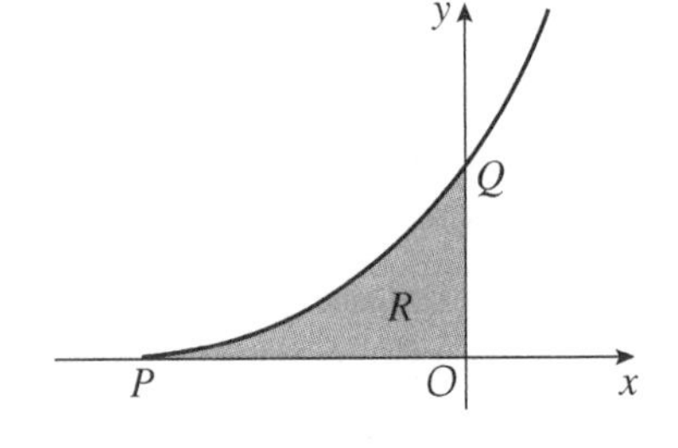

(a) Find the exact coordinates of the points P and Q.

(b) Show that $y\dfrac{dx}{dt} = 2t^4$.

(c) Hence find the area of R.

11 The diagram shows a sketch of the curve with parametric

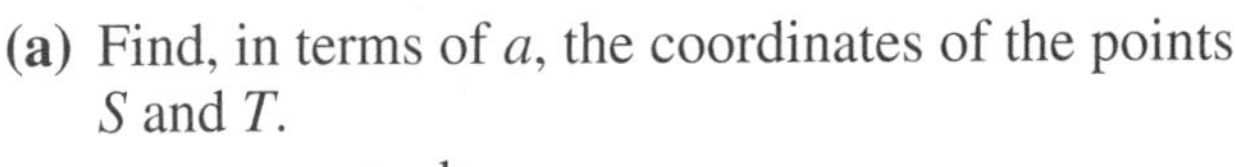

equations $x = at$, $y = \dfrac{4a}{t}$, $t > 0$, and the line $y = 5a - x$, where a is a constant. The line meets the curve at S and T.

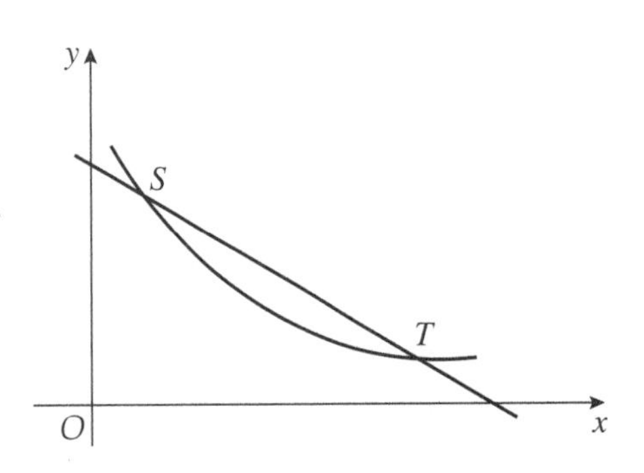

(a) Find, in terms of a, the coordinates of the points S and T.

(b) Show that $\displaystyle\int y\frac{dx}{dt}\,dt = 4a^2\ln t + c$, where c is a constant.

(c) Hence find, in terms of a, the exact area of the finite region between the curve and the line.

12 The diagram shows a sketch of the curve with parametric equations $x = 8 - t^2$, $y = 2t$, $t > 0$.
The curve meets the y-axis at P.

(a) Find the exact value of t at the point P.

(b) Find the area bounded by the curve and the positive coordinate axes.

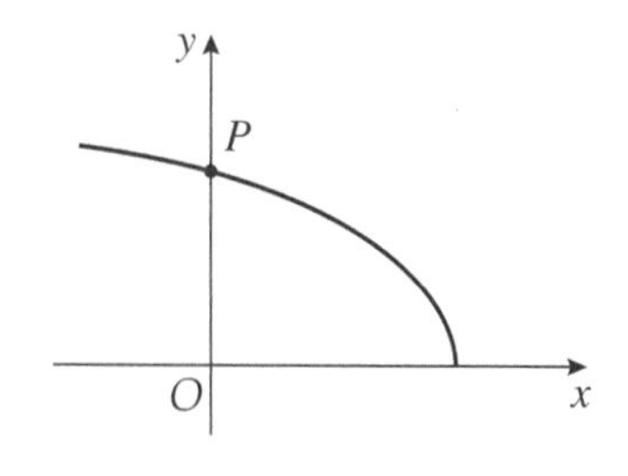

13 The diagram shows a sketch of the curve with parametric equations $x = 1 - t^2$, $y = t(4 - t^2)$.
The curve meets the x-axis at A and B.

(a) Find the values of t at the points A and B.

(b) Show that $y\dfrac{dx}{dt}\,dt = 2t^4 - 8t^2$.

(c) Hence, find the area enclosed by the loop of the curve.

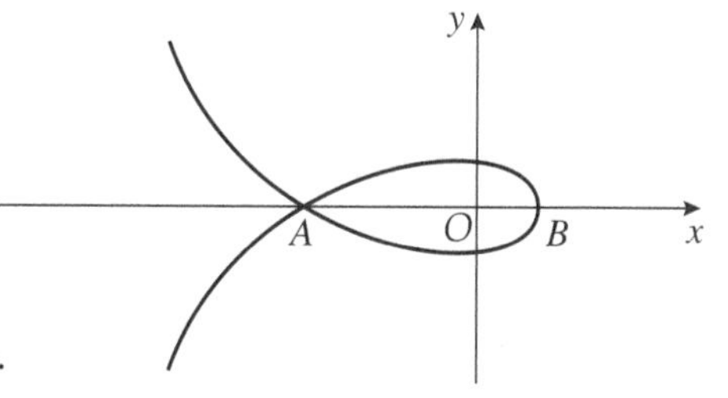

14 The diagram shows a sketch of the cycloid with parametric equations $x = \theta - \sin\theta$, $y = 1 - \cos\theta$, $0 \leqslant \theta \leqslant 2\pi$.
The cycloid meets the x-axis at 0 and P.

(a) Find the exact coordinates of P.

The area enclosed by the cycloid and the x-axis is A, where $A = \displaystyle\int_0^{2\pi} y\frac{dx}{d\theta}\,d\theta$.

(b) Show that $A = \displaystyle\int_0^{2\pi} (1 - \cos\theta)^2\,d\theta$.

(c) Hence, find the exact value of A.

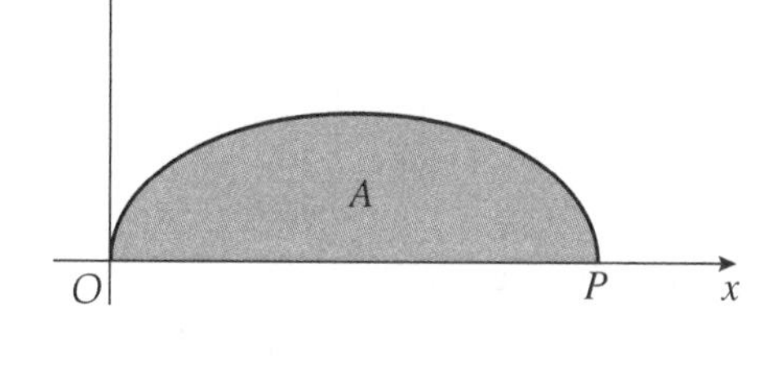

15 The diagram shows a sketch of the curve with parametric equations $x = 4\sin\theta$, $y = 3\sin 2\theta$, $0 \leqslant \theta \leqslant 2\pi$.
The curve is symmetrical in both axes. The shaded region R is bounded by the curve, $0 \leqslant \theta \leqslant \frac{\pi}{2}$, and the x-axis.

(a) Find the area of R.

(b) Write down the total area enclosed by the curve.

(c) Show that the Cartesian equation of the curve can be written in the form $y^2 = ax^2(16 - x^2)$, where a is a constant to be found.

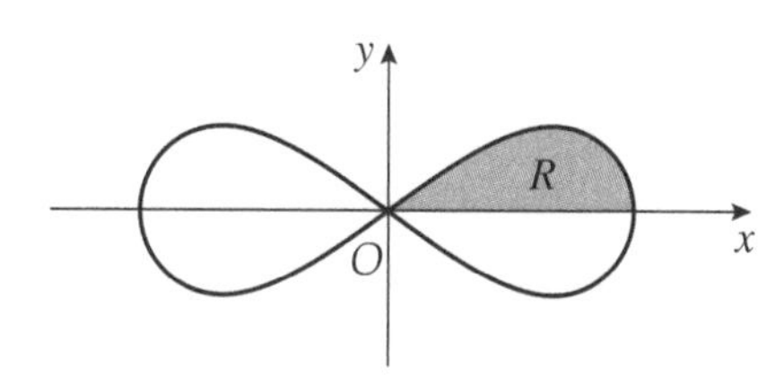

16 The diagram shows part of the curve with parametric equations $x = \tan t$, $y = \sin 2t$, $-\frac{\pi}{2} < t < \frac{\pi}{2}$.

(a) Find the gradient of the curve at the point P where $t = \frac{\pi}{3}$.

(b) Find an equation of the normal to the curve at P.

(c) Find an equation of the normal to the curve at the point Q where $t = \frac{\pi}{4}$.

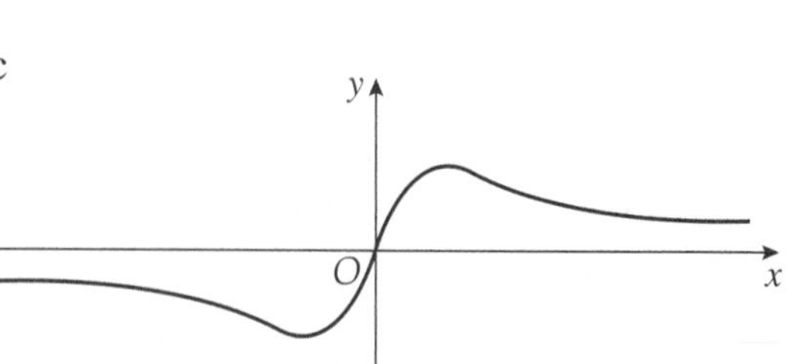

The binomial expansion 3

What you should know

1 The binomial expansion $(1 + x)^n = 1 + nx + \frac{n(n-1)x^2}{2!} + \frac{n(n-1)(n-2)x^3}{3!} + \ldots$

can be used to express an exact expression if n is a positive integer, or an approximate expression for any other rational number.

- $(1 + 2x)^3 = 1 + 3(2x) + 3 \times 2\frac{(2x)^2}{2!} + 3 \times 2 \times 1 \times \frac{(2x)^3}{3!} + 3 \times 2 \times 1 \times 0 \times \frac{(2x)^4}{4!}$

  $= 1 + 6x + 12x^2 + 8x^3$ (Expansion is *finite* and *exact*)

- $\sqrt{(1-x)} = (1-x)^{\frac{1}{2}} = 1 + \left(\frac{1}{2}\right)(-x) + \left(\frac{1}{2}\right)\left(-\frac{1}{2}\right)\frac{(-x)^2}{2!} + \left(\frac{1}{2}\right)\left(-\frac{1}{2}\right)\left(-\frac{3}{2}\right)\frac{(-x)^3}{3!} + \ldots$

  $= 1 - \frac{1}{2}x - \frac{1}{8}x^2 - \frac{1}{16}x^3 - \ldots$

  (Expansion is *infinite* and *approximate*)

2 The expansion $(1 + x)^n = 1 + nx + n(n-1)\frac{x^2}{2!} + n(n-1)(n-2)\frac{x^3}{3!} + \ldots,$

where n is negative or a fraction, is only valid if $|x| < 1$.

3 You can adapt the binomial expansion to include expressions of the form $(a + bx)^n$ by simply taking out a common factor of a:

e.g. $\frac{1}{(3+4x)} = (3 + 4x)^{-1} = \left[3\left(1 + \frac{4x}{3}\right)\right]^{-1}$

$= 3^{-1}\left(1 + \frac{4x}{3}\right)^{-1}$

4 You can use knowledge of partial fractions to expand more difficult expressions, e.g.

$$\frac{7+x}{(3-x)(2+x)} = \frac{2}{(3-x)} + \frac{1}{(2+x)}$$

$$= 2(3-x)^{-1} + (2+x)^{-1}$$

$$= \frac{2}{3}\left(1 - \frac{x}{3}\right)^{-1} + \frac{1}{2}\left(1 + \frac{x}{2}\right)^{-1}$$

Test yourself

What to review

If your answer is incorrect

1 Use the binomial expansion to find, in ascending powers of x, the first four terms of $\frac{1}{(1+3x)^2}$. State the range in values of x for which the expansion is valid.

Review Edexcel Book C4 pages 24–27
Revise for C4 page 23 Example 1

---

2 Find, in ascending powers of x, the first four terms in the binomial expansions of

(a) $\sqrt{9-x}$ **(b)** $(3+x)\sqrt{9-x}$.

State the range in values of x for which the expansions are valid.

Review Edexcel Book C4 pages 29–30
Revise for C4 page 24 Example 2 and page 25 Worked examination style question 1

---

3 In the binomial expansion of $(4+bx)^{\frac{1}{2}}$ the coefficient of x^2 is -9. Find:

(a) the possible values of b,

(b) the corresponding coefficient of x.

Review Edexcel Book C4 pages 29–30
Revise for C4 page 25 Example 3

---

4 **(a)** Show that if $|2x| < 1$, then $\sqrt{1-2x} \approx 1 - x - \frac{x^2}{2}$.

(b) By substituting $x = 0.05$, find an approximation to $\sqrt{10}$ giving your answer to 3 significant figures.

Review Edexcel Book C4 pages 27–28
Revise for C4 page 28 Worked examination style question 3

---

5 Given that $\frac{10x}{(2+x)(3-x)} \equiv \frac{A}{(2+x)} + \frac{B}{(3-x)}$

(a) find the values of the constants A and B.

(b) Hence, or otherwise, find the binomial expansion of $\frac{10x}{(2+x)(3-x)}$ in ascending powers of x up to and including the term in x^3.

(c) State the range in values for which the expansion in **(b)** is valid.

Review Edexcel Book C4 pages 31–33
Revise for C4 page 27 Worked examination style question 2

Example 1

Use the binomial expansion to find the first four terms of $\sqrt{1+2x}$.
State the range in values of x for which this expansion is valid.

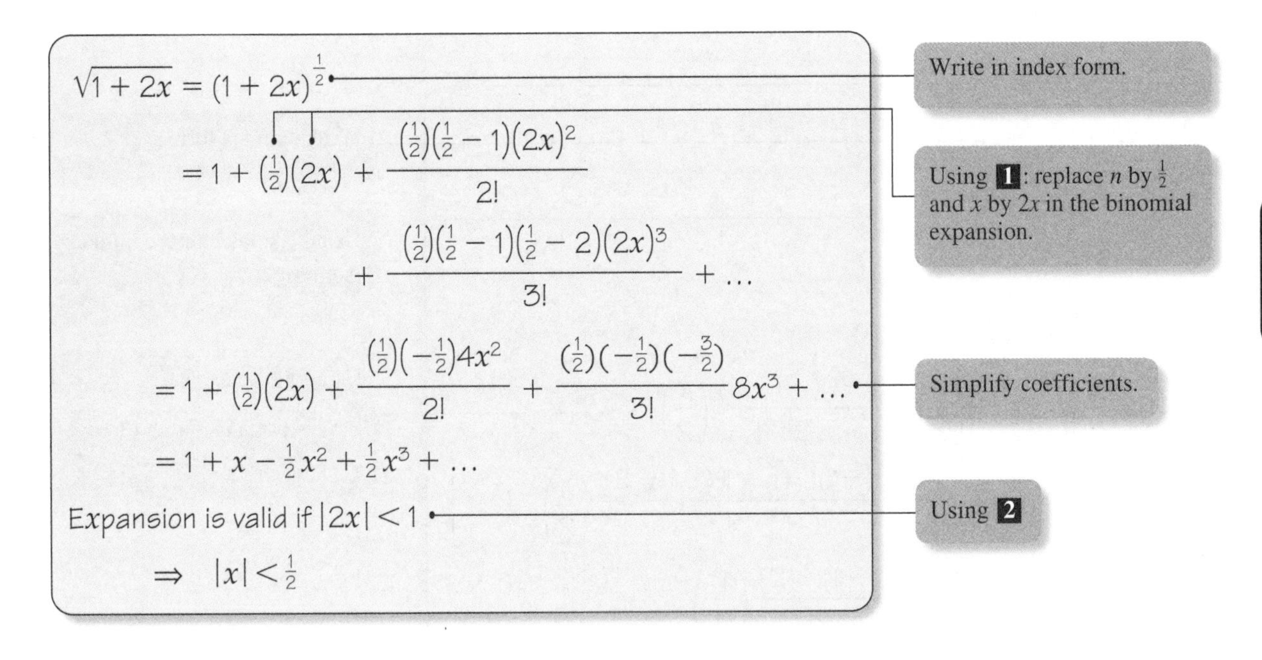

Example 2

Find the expansion of $\dfrac{1}{2+x}$ in ascending powers of x as far as x^3.
State the range in values of x for which this expansion is valid.

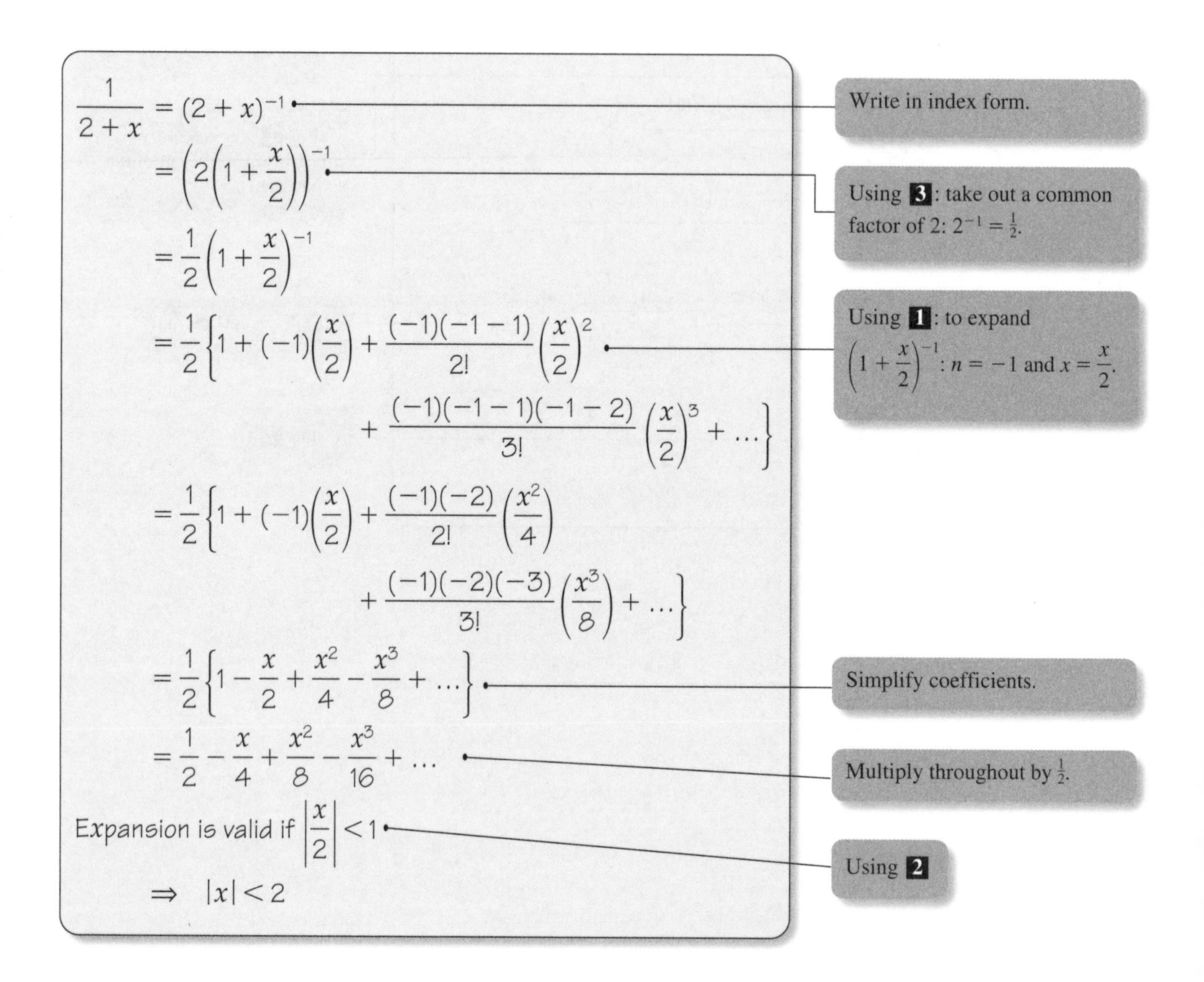

$$\frac{1}{2+x} = (2+x)^{-1}$$

Write in index form.

$$= \left(2\left(1+\frac{x}{2}\right)\right)^{-1}$$

Using **3**: take out a common factor of 2: $2^{-1} = \frac{1}{2}$.

$$= \frac{1}{2}\left(1+\frac{x}{2}\right)^{-1}$$

$$= \frac{1}{2}\left\{1 + (-1)\left(\frac{x}{2}\right) + \frac{(-1)(-1-1)}{2!}\left(\frac{x}{2}\right)^2 + \frac{(-1)(-1-1)(-1-2)}{3!}\left(\frac{x}{2}\right)^3 + \dots\right\}$$

Using **1**: to expand $\left(1+\frac{x}{2}\right)^{-1}$: $n = -1$ and $x = \frac{x}{2}$.

$$= \frac{1}{2}\left\{1 + (-1)\left(\frac{x}{2}\right) + \frac{(-1)(-2)}{2!}\left(\frac{x^2}{4}\right) + \frac{(-1)(-2)(-3)}{3!}\left(\frac{x^3}{8}\right) + \dots\right\}$$

$$= \frac{1}{2}\left\{1 - \frac{x}{2} + \frac{x^2}{4} - \frac{x^3}{8} + \dots\right\}$$

Simplify coefficients.

$$= \frac{1}{2} - \frac{x}{4} + \frac{x^2}{8} - \frac{x^3}{16} + \dots$$

Multiply throughout by $\frac{1}{2}$.

Expansion is valid if $\left|\frac{x}{2}\right| < 1$

Using **2**

$$\Rightarrow \quad |x| < 2$$

Example 3

In the expansion of $(1 + bx)^{\frac{3}{2}}$ the coefficient of x^2 is $\frac{2}{3}$.
Find possible values of the constant b.

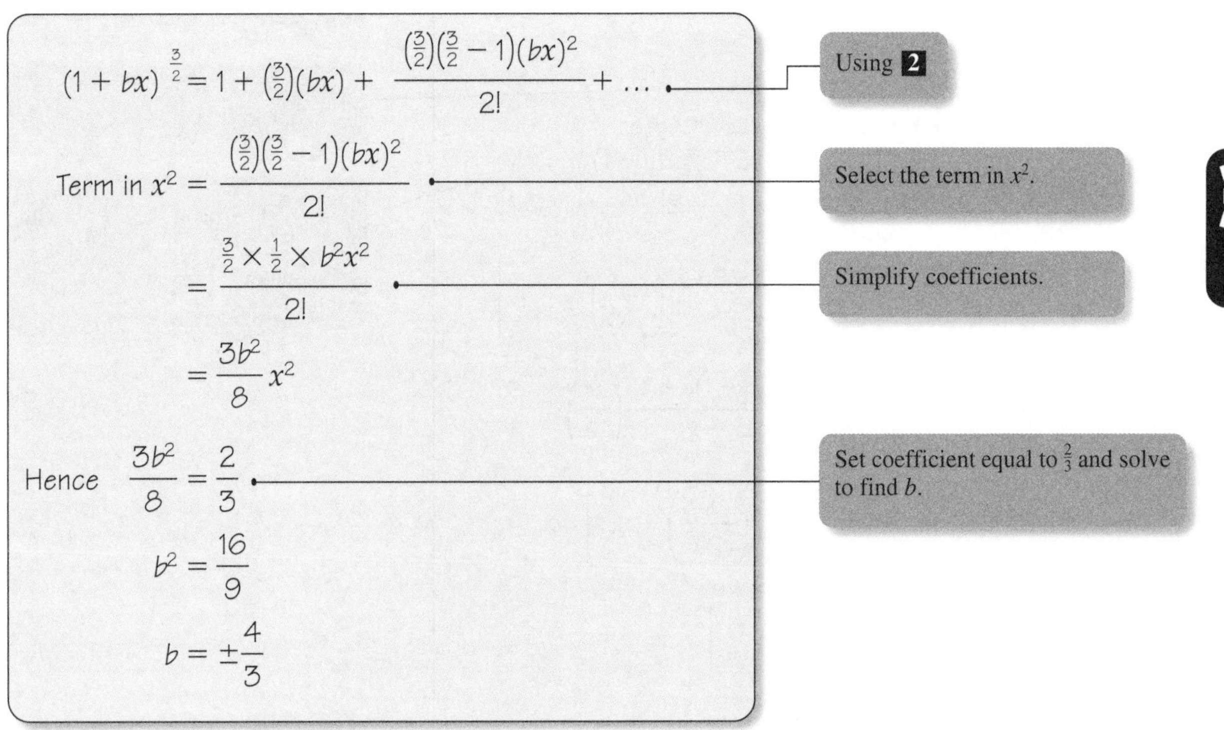

Worked examination style question 1

Given that $|x| < 3$, find, in ascending powers of x up to and including the term in x^3, the series expansion of:

(a) $(3 - x)^{-1}$

(b) $\frac{(1 + 2x)}{3 - x}$.

(a) $(3 - x)^{-1}$

$$= \left(3\left(1 - \frac{x}{3}\right)\right)^{-1}$$

Using **3**: take 3 out as a common factor.

$$= \frac{1}{3}\left(1 - \frac{x}{3}\right)^{-1}$$

$3^{-1} = \frac{1}{3}$.

$$= \frac{1}{3}\left[1 + (-1)\left(-\frac{x}{3}\right) + \frac{(-1)(-1-1)}{2!}\left(-\frac{x}{3}\right)^2 + \frac{(-1)(-1-1)(-1-2)}{3!}\left(-\frac{x}{3}\right)^3 + \ldots\right]$$

Using **2**: expand $\left(1 - \frac{x}{3}\right)^{-1}$ with $n = 1$ and $x = -\frac{x}{3}$.

$$= \frac{1}{3}\left[1 + (-1)\left(-\frac{x}{3}\right) + \frac{(-1)(-2)}{2!}\left(\frac{x^2}{9}\right) + \frac{(-1)(-2)(-3)}{3!}\left(\frac{-x^3}{27}\right) + \ldots\right]$$

$$= \frac{1}{3}\left[1 + \frac{x}{3} + \frac{x^2}{9} + \frac{x^3}{27} + \ldots\right]$$

$$= \frac{1}{3} + \frac{x}{9} + \frac{x^2}{27} + \frac{x^3}{81} + \ldots$$

Multiply by $\frac{1}{3}$.

(b) $\dfrac{(1 + 2x)}{3 - x}$

$$= (1 + 2x)(3 - x)^{-1}$$

Write $\frac{1}{3 - x}$ as $(3 - x)^{-1}$.

$$= (1 + 2x)\left(\frac{1}{3} + \frac{x}{9} + \frac{x^2}{27} + \frac{x^3}{81} + \ldots\right)$$

Use **2** for the expansion of $(3 - x)^{-1}$.

$$= \frac{1}{3} + \frac{x}{9} + \frac{x^2}{27} + \frac{x^3}{81} + \frac{2x}{3} + \frac{2x^2}{9} + \frac{2x^3}{27} + \ldots$$

Expand the brackets
Omit terms of x^4 and above.

$$= \frac{1}{3} + \frac{7}{9}x + \frac{7}{27}x^2 + \frac{7}{81}x^3 + \ldots$$

Collect like terms.

Worked examination style question 2

When $(1 + ax)^n$ is expanded as a series in ascending powers of x the coefficients of x and x^2 are -8 and 48 respectively.

(a) Find the values of a and n.

(b) Find the coefficients of x^3.

(c) State the values of x for which the expansion is valid.

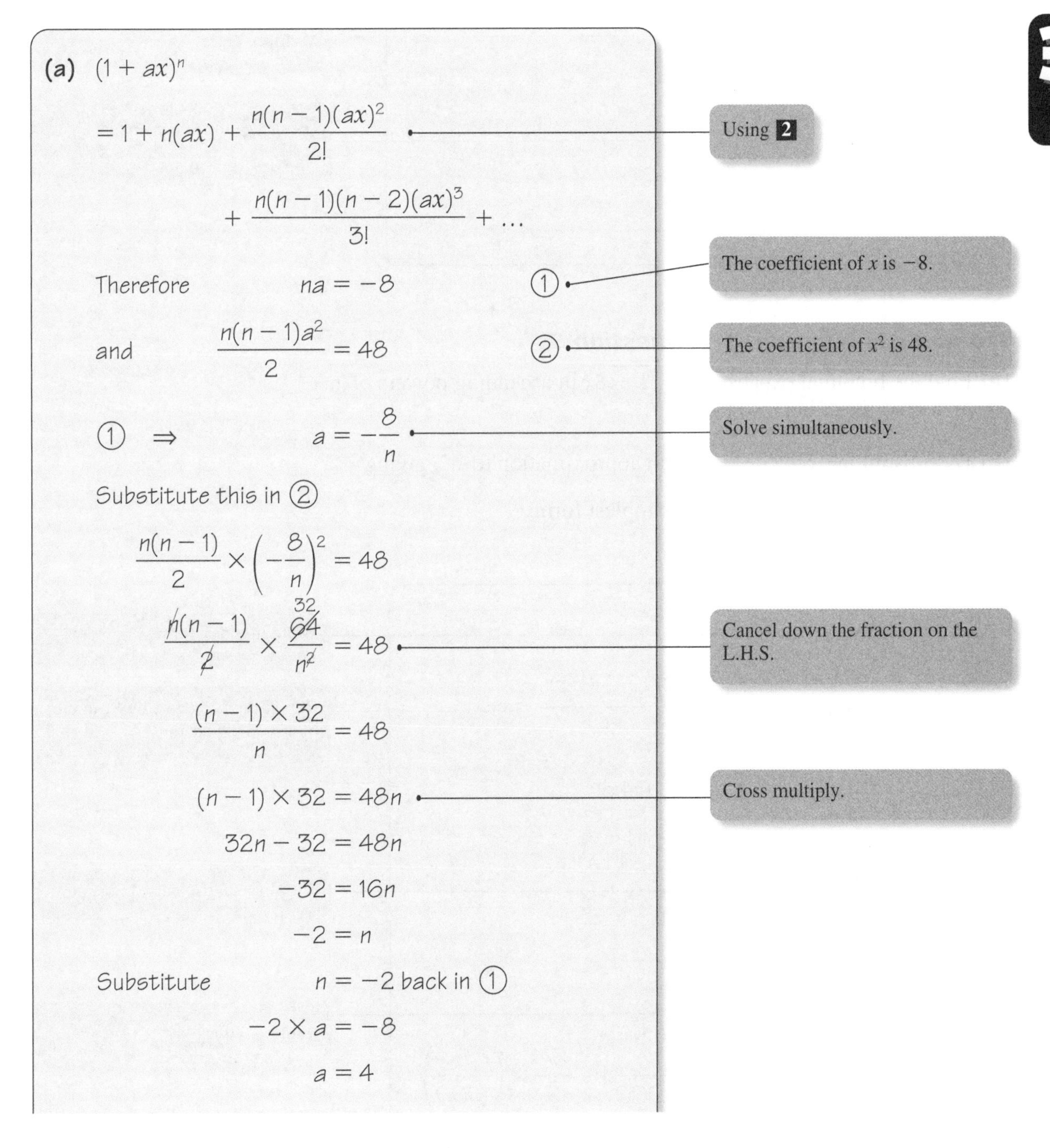

(a) $(1 + ax)^n$

$$= 1 + n(ax) + \frac{n(n-1)(ax)^2}{2!} + \frac{n(n-1)(n-2)(ax)^3}{3!} + \dots$$

Using **2**

Therefore $na = -8$ ①

The coefficient of x is -8.

and $\frac{n(n-1)a^2}{2} = 48$ ②

The coefficient of x^2 is 48.

① $\Rightarrow$ $a = -\frac{8}{n}$

Solve simultaneously.

Substitute this in ②

$$\frac{n(n-1)}{2} \times \left(-\frac{8}{n}\right)^2 = 48$$

$$\frac{n(n-1)}{2} \times \frac{64}{n^2} = 48$$

Cancel down the fraction on the L.H.S.

$$\frac{(n-1) \times 32}{n} = 48$$

$$(n-1) \times 32 = 48n$$

Cross multiply.

$$32n - 32 = 48n$$

$$-32 = 16n$$

$$-2 = n$$

Substitute $n = -2$ back in ①

$$-2 \times a = -8$$

$$a = 4$$

(b) The x^3 term is $\dfrac{n(n-1)(n-2)(ax)^3}{3!}$

$= \dfrac{-2(-2-1)(-2-2)(4x)^3}{3!}$

Substitute $a = 4$ and $n = -2$ in the x^3 term.

$= \dfrac{-2 \times -3 \times -4 \times 64x^3}{3!}$

Simplify brackets.

$= -256x^3$

Multiply out.

The coefficient of x^3 is -256

(c) Expansion is valid if $|ax| < 1$

Using **2**: the expansion is $(1 + ax)^n$.

$|4x| < 1$

$|x| < \frac{1}{4}$

Worked examination style question 3

(a) Find the binomial expansion of $\sqrt[3]{1 + 8x}$ in ascending powers of x, up to and including the term in x^2.

(b) By substituting $x = \dfrac{3}{1000}$, find an approximation to $\sqrt[3]{2}$, giving your answer as a fraction in its simplest form.

(a) $\sqrt[3]{1 + 8x} = (1 + 8x)^{\frac{1}{3}}$

Write in index form.

$= 1 + (\frac{1}{3})(8x) + \dfrac{(\frac{1}{3})(\frac{1}{3} - 1)(8x)^2}{2!}$

Replace n by $\frac{1}{3}$ and x by $8x$ in the binomial expansion.

$= 1 + \dfrac{8x}{3} + \dfrac{(\frac{1}{3})(-\frac{2}{3})64x^2}{2!}$

$= 1 + \dfrac{8x}{3} - \dfrac{64x^2}{9}$

Simplify coefficients.

(b) Substitute $x = \dfrac{3}{1000}$

$\sqrt[3]{1 + 8 \times \dfrac{3}{1000}} = 1 + \dfrac{8}{3} \times \left(\dfrac{3}{1000}\right) - \dfrac{64}{9} \times \left(\dfrac{3}{1000}\right)^2$

Replace x by $\dfrac{3}{1000}$.

$$\sqrt[3]{1 + \frac{24}{1000}} = 1 + \frac{8 \times \cancel{3}}{\cancel{3} \times 1000} - \frac{64 \times \cancel{9}}{\cancel{9} \times 1\,000\,000}$$

Simplify terms.

$$\sqrt[3]{\frac{1024}{1000}} = 1 + \frac{8}{1000} - \frac{64}{1\,000\,000}$$

Write both sides as improper fractions.

$$\sqrt[3]{\frac{1024}{1000}} = \frac{1\,007\,936}{1\,000\,000}$$

Write 1024 as 2×512.

$$\sqrt[3]{\frac{512 \times 2}{1000}} = \frac{1\,007\,936}{1\,000\,000}$$

$\sqrt[3]{512} = 8$ and $\sqrt[3]{1000} = 10$.

$$\frac{8 \times \sqrt[3]{2}}{10} = \frac{1\,007\,936}{1\,000\,000}$$

$$\sqrt[3]{2} = \frac{1\,007\,936}{1\,000\,000} \times \frac{10}{8}$$

Multiply both sides by $\frac{10}{8}$.

$$\sqrt[3]{2} = \frac{125\,992}{100\,000} = \frac{15\,749}{12\,500}$$

Simplify the right hand side.

Revision exercise 3

1 Find the binomial expansions of the following, in ascending powers of x as far as the term in x^3. State the set of values of x for which the expansion is valid.

(a) $(1 + x)^{\frac{1}{2}}$ **(b)** $\frac{1}{(1 - 3x)^2}$ **(c)** $\frac{1}{\sqrt{4 - x}}$

2 Given that $|x| < \frac{1}{2}$, find the first four terms in ascending powers of x in the expansion of:

(a) $(1 - 2x)^{\frac{1}{2}}$ **(b)** $(3 + x)(1 - 2x)^{\frac{1}{2}}$.

3 Given that $|x| < 2$, find the first four terms in ascending powers of x in the expansion of:

(a) $\frac{1}{2 - x}$ **(b)** $\frac{x^2 + 3x - 1}{2 - x}$.

4 In the expansion of $(1 + ax)^{-2}$ the coefficient of x^2 is 12. Find possible values of the constant a and hence the corresponding values of the term in x^3.

5 Show that if x is small, the expression $\dfrac{3+x}{1-x}$ can be approximated by the quadratic expression $3 + 4x + 4x^2$.

6 Find the binomial expansion of $\dfrac{1}{\sqrt{1-x}}$, in ascending powers of x up to and including the term in x^2. By substituting $x = \frac{1}{4}$ in this expansion, find a fractional approximation to $\sqrt{3}$.

7 Obtain the first four non-zero terms in ascending powers of x in the expansion of $\dfrac{1}{\sqrt{1-2x^2}}$, $2x^2 < 1$.

8 When $(1 + ax)^n$ is expanded as a series, in ascending powers of x, the coefficients of x and x^2 are 3 and $-4\frac{1}{2}$ respectively.

(a) Find the values of a and n.

(b) Find the coefficient of x^3.

(c) State the values of x for which the expansion is valid.

9 **(a)** Express $\dfrac{11+4x}{(4-x)(1+2x)}$ in the form $\dfrac{A}{(4-x)} + \dfrac{B}{(1+2x)}$ where A and B are constants to be determined.

(b) Hence or otherwise find the binomial expansion, in ascending powers of x, of $\dfrac{11+4x}{(4-x)(1+2x)}$. Expand as far as the term in x^2.

(c) State the set of values of x for which the expansion is valid.

10 Given that $\dfrac{11x-1}{(1-x)^2(2+3x)} \equiv \dfrac{A}{(1-x)^2} + \dfrac{B}{(1-x)} + \dfrac{C}{(2+3x)}$

(a) find the values of A, B and C.

(b) Hence or otherwise find the series expansion of $\dfrac{11x-1}{(1-x)^2(2+3x)}$ in ascending powers of x up to and including the term in x^2.

(c) State the values of x for which the expansion in **(b)** is valid.

11 Use the binomial theorem to expand $(4 - 3x)^{-\frac{1}{2}}$, in ascending powers of x, up to and including the term in x^3.
Give each coefficient as a simplified fraction. **E**

Differentiation

What you should know

1 When a relation is described by parametric equations:

- You differentiate x and y with respect to the parameter t.
- Then you use the chain rule rearranged into the form $\dfrac{dy}{dx} = \dfrac{dy}{dt} \div \dfrac{dx}{dt}$.

2 When a relation is described by an *implicit* equation:

- Differentiate each term in turn, using the chain rule and product and quotient rules as appropriate.
- $\dfrac{d}{dx}(y^n) = ny^{n-1}\dfrac{dy}{dx}$ — By the chain rule
- $\dfrac{d}{dx}(xy) = x\dfrac{dy}{dx} + y$ — By the product rule

3 In an implicit equation:

- Note that when $f(y)$ is differentiated with respect to x it becomes $f'(y)\dfrac{dy}{dx}$.
- A product term such as $f(x)g(y)$ is differentiated by the product rule and becomes $f(x)g'(y)\dfrac{dy}{dx} + g(y)f'(x)$.

4 You can differentiate the function $f(x) = a^x$, where a is constant:

- If $y = a^x$, then $\dfrac{dy}{dx} = a^x \ln a$.

5 You can use the chain rule once, or several times to connect the rates of change in a question involving more than two variables.

6 You can set up simple differential equations from information given in context. This may involve using connected rates of change, or ideas of proportion.

Test yourself

	What to review
	If your answer is incorrect
1 Giving your answer in terms of t, find the gradient of the curve with parametric equations: **(a)** $x = t^2$, $y = 2t^3$ **(b)** $x = \mathrm{e}^t + \mathrm{e}^{-t}$, $y = \mathrm{e}^t - \mathrm{e}^{-t}$ **(c)** $x = \sin 2t$, $y = \sin t$.	*Review Edexcel Book C4 page 37* *Revise for C4 page 33 Example 1 and page 34 Worked examination style question 1*
2 For the curve with equation $x = \cos 2t$, $y = \tan t$, find the equation of the: **(a)** tangent to the curve at the point where $t = \frac{\pi}{3}$, **(b)** normal to the curve at the point where $t = \frac{\pi}{4}$.	*Review Edexcel Book C4 pages 37–38* *Revise for C4 page 33 Example 1 and page 34 Worked examination style question 1*
3 Find $\frac{\mathrm{d}y}{\mathrm{d}x}$ in terms of x and y where: **(a)** $x^2 + 3y^2 - 6x = 12$ **(b)** $x^4 + 4x^2y^2 + y = 8$.	*Review Edexcel Book C4 pages 39–40* *Revise for C4 page 33 Example 2 and page 35 Worked examination style question 3*
4 Find $\frac{\mathrm{d}y}{\mathrm{d}x}$ when: **(a)** $y = 2^x$ **(b)** $y = x3^x$ **(c)** $y = 4^{\sqrt{x}}$.	*Review Edexcel Book C4 page 41* *Revise for C4 page 34 Example 4*
5 The radius r cm of a sphere is expanding at a rate, measured in cm s^{-1}, which is inversely proportional to r. When $r = 1$, the rate of increase is 4 cm s^{-1}. The volume of the sphere is V cm^3, where $V = \frac{4}{3}\pi r^3$ and the surface area is S cm^2, where $S = 4\pi r^2$. Find: **(a)** $\frac{\mathrm{d}V}{\mathrm{d}t}$ **(b)** $\frac{\mathrm{d}S}{\mathrm{d}t}$.	*Review Edexcel Book C4 page 42* *Revise for C4 page 33 Example 3 and page 35 Worked examination style question 2*
6 A curve C has equation $y = \mathrm{f}(x)$ and its gradient at each point on the curve is directly proportional to the square of its y-coordinate at that point. At the point (0, 3) on the curve the gradient is 1. Write down a differential equation, which could be solved to give the equation of the curve. State the value of any constant of proportionality which you use.	*Review Edexcel Book C4 pages 43–45*

Example 1

Find the gradient of the curve with parametric equations $x = \dfrac{t^2}{2t+1}$, $y = \dfrac{1}{2t+1}$, at the point $\left(\frac{1}{8}, \frac{1}{2}\right)$.

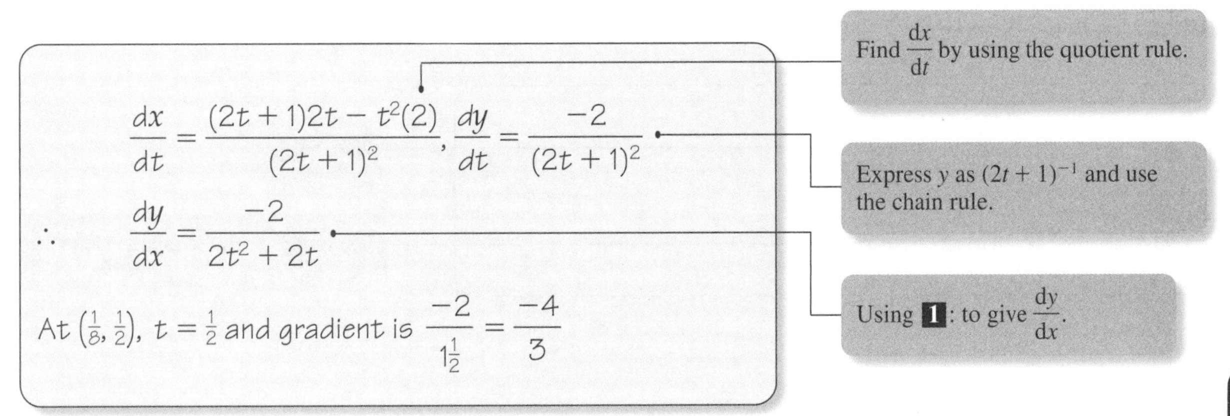

$$\frac{dx}{dt} = \frac{(2t+1)2t - t^2(2)}{(2t+1)^2}, \quad \frac{dy}{dt} = \frac{-2}{(2t+1)^2}$$

$$\therefore \quad \frac{dy}{dx} = \frac{-2}{2t^2 + 2t}$$

At $\left(\frac{1}{8}, \frac{1}{2}\right)$, $t = \frac{1}{2}$ and gradient is $\dfrac{-2}{1\frac{1}{2}} = \dfrac{-4}{3}$

Find $\dfrac{dx}{dt}$ by using the quotient rule.

Express y as $(2t+1)^{-1}$ and use the chain rule.

Using **1**: to give $\dfrac{dy}{dx}$.

Example 2

Use implicit differentiation to find $\dfrac{dy}{dx}$ in terms of x and y where $x^4 - 6x^2y + 4y^3 = 2$.

$$4x^3 - 12xy - 6x^2\frac{dy}{dx} + 12y^2\frac{dy}{dx} = 0$$

$$\therefore \quad (12y^2 - 6x^2)\frac{dy}{dx} = 12xy - 4x^3$$

$$\therefore \quad \frac{dy}{dx} = \frac{12xy - 4x^3}{12y^2 - 6x^2} = \frac{2x(3y - x^2)}{3(2y^2 - x^2)}$$

Using **2**: differentiate term by term.

Collect $\dfrac{dy}{dx}$ terms, factorise and make $\dfrac{dy}{dx}$ the subject of the formula.

Example 3

Given that the volume V cm^3 of a hemisphere is related to its radius r cm by the formula $V = \frac{2}{3}\pi r^3$ and that the hemisphere expands so that the rate of increase of its radius in cm s^{-1} is given by $\dfrac{dr}{dt} = \dfrac{2}{r^3}$, find the exact value of $\dfrac{dV}{dt}$ when $r = 3$.

$$\frac{dV}{dr} = 2\pi r^2$$

$$\frac{dV}{dt} = \frac{dV}{dr} \times \frac{dr}{dt}$$

$$\therefore \quad \frac{dV}{dt} = 2\pi r^2 \times \frac{2}{r^3} = \frac{4\pi}{r}$$

When $r = 3$ the value of $\dfrac{dV}{dt} = \dfrac{4\pi}{3}$

Using **5**: to connect rates of change.

Example 4

Find $\frac{dy}{dx}$ when:

(a) $y = 10^x$ **(b)** $y = 5^{x^2}$.

(a) $\frac{dy}{dx} = \ln 10 \times 10^x$

Using **4**

(b) $\frac{dy}{dx} = 2x \times \ln 5 \times 5^{x^2}$

Using **4**: with the chain rule.

Example 5

The rate of increase of a population is proportional to the size, P, of the population. Express this as a differential equation.

$\frac{dP}{dt} = kP$

Using **6**: k is a constant of proportionality.

Worked examination style question 1

The curve C is given by the equations $y = 2t$, $x = t^2 + t^3$ where t is a parameter.

Find the equation of the normal to C at the point P on C where $t = -2$.

$\frac{dy}{dt} = 2, \frac{dx}{dt} = 2t + 3t^2$

$\frac{dy}{dx} = \frac{2}{2t + 3t^2}$

Using **1**

The gradient of the curve at the point where $t = -2$ is $\frac{2}{8} = \frac{1}{4}$

The gradient of the normal is -4

The normal is perpendicular to the curve so its gradient is $-1 \div m$, where m is the gradient of the curve.

When $t = -2$, $x = -4$ and $y = -4$

The equation of the normal is $y - (-4) = -4(x - (-4))$
i.e. $y + 4x + 20 = 0$

Using: $y - y_1 = m(x - x_1)$.

Worked examination style question 2

At time t seconds the surface area of a cube is A cm^2 and its volume is V cm^3. The volume of the cube is expanding at a uniform rate of 2 cm^3 s^{-1}.

Show that $\frac{dA}{dt} = kA^{-\frac{1}{2}}$, where k is a constant to be determined.

If the side of the cube is x cm then $V = x^3$

By differentiation $\frac{dV}{dx} = 3x^2$ and $\frac{dV}{dt} = 2$ (given)

So by the chain rule $\frac{dx}{dt} = \frac{2}{3x^2}$

Using **5**: $\frac{dx}{dt} \times \frac{dV}{dx} = \frac{dV}{dt}$

Also $A = 6x^2$, so $\frac{dA}{dx} = 12x$ and so $\frac{dA}{dt} = 12x \times \frac{2}{3x^2} = \frac{8}{x}$

$\frac{dA}{dt} = \frac{dA}{dx} \times \frac{dx}{dt}$

As $x = \sqrt{\frac{A}{6}}$, $\frac{dA}{dt} = \frac{8\sqrt{6}}{\sqrt{A}}$ and so $k = 8\sqrt{6}$

Use $A = 6x^2$ and make x the subject of the formula.

Worked examination style question 3

Given that $y^2(x^2 + xy) = 12$, evaluate $\frac{dy}{dx}$ when $y = 1$.

When $y = 1$, $x^2 + x = 12$

Substitute $y = 1$ to obtain values for x.

$\therefore \quad x = 3$ or -4

Differentiate to give $y^2\left(2x + y + x\frac{dy}{dx}\right) + 2y\frac{dy}{dx}(x^2 + xy) = 0$

Using **2**: to differentiate term by term.

When $x = 3$, $y = 1$, $1\left(7 + 3\frac{dy}{dx}\right) + 2\frac{dy}{dx}(12) = 0$

$\therefore \quad \frac{dy}{dx} = -\frac{7}{27}$

When $x = -4$, $y = 1$, $1\left(-7 - 4\frac{dy}{dx}\right) + 2\frac{dy}{dx}(12) = 0$

$\therefore \quad \frac{dy}{dx} = \frac{7}{20}$

Revision exercise 4

1 The curve C is described by the parametric equations
$x = 3 \cos t$, $y = \cos 2t$, $0 < t < \pi$.
Find the coordinates of the point on the curve where the gradient is 1.

2 The ellipse has parametric equations
$x = 5 \cos \theta$, $y = 4 \sin \theta$, $0 \leqslant \theta < 2\pi$.
Show that an equation of the tangent to the ellipse at $(5 \cos \alpha, 4 \sin \alpha)$, is $5y \sin \alpha + 4x \cos \alpha = 20$.

3 The curve C has parametric equations $y = 3^t$, $x = 3t$, $t \in \mathbb{R}$.
Find the equation of the tangent to C at the point where $t = 0$.

4 The curve C has equation $5x^2 + 2xy - 3y^2 + 3 = 0$.
The point P on the curve C has coordinates $(1, 2)$.
Find the gradient of the curve at P.
Find the equation of the normal to the curve at P in the form $y = ax + b$, where a and b are constants.

5 A curve has equation $7x^2 + 48xy - 7y^2 + 75 = 0$.
A and B are two distinct points on the curve.
At each of these points the gradient of the curve is equal to $\frac{2}{11}$.
Use implicit differentiation to show that $x + 2y = 0$ at the points A and B.

6 Find the coordinates of the turning points on the curve with equation $y^3 + 3xy^2 - x^3 = 3$. *E*

7 Given that $e^{2x} + e^{2y} = xy$, find $\frac{dy}{dx}$ in terms of x and y. *E*

8 Find the gradient of the curve with equation
$x^3 + y^3 = 2(x^2 + y^2)$ at the point with coordinates $(2, 2)$.

9 A drop of oil is modelled as a circle of radius r.
At time t, $t > 0$, $r = 4(1 - e^{-\lambda t})$, where λ is a positive constant.
Show that the area A of the circle satisfies

$$\frac{dA}{dt} = 32\pi\lambda(e^{-\lambda t} - e^{-2\lambda t}).$$

E

10 At time t the rate of increase in the concentration x of micro-organisms in controlled surroundings, is equal to k times the concentration, where k is a positive constant.
Write down a differential equation in x, t and k. *E*

11 A piece of radioactive material is disintegrating. At time t days after it was first observed its mass is m grams. It is noted that the rate of loss of mass is proportional to the mass remaining. Express this as a differential equation.

12 Newton's law of cooling states that the rate of loss of temperature of a body is directly proportional to the excess temperature of the body over the temperature of its surroundings.
Given that at time t minutes a body has temperature T °C and its surroundings are at a constant temperature of 22 °C, form a differential equation in terms of t, T and k, where k is a constant of proportionality greater than zero.

13 Fluid flows out of a cylindrical tank with a constant cross-section. At time t, the volume of fluid remaining in the tank is V.
The rate at which the fluid flows is proportional to the square root of V.
Show that the depth h of fluid in the tank satisfies the differential equation $\frac{dh}{dt} = -k\sqrt{h}$, where k is a positive constant. **E**

14 The rate of cooling of a metal ball placed in melting ice is proportional to its own temperature T °C.
Show, by differentiating, that at time t, $T = Ae^{-kt}$, where A and k are positive constants.

15 The population, p, of insects on an island, t hours after midday is given by $p = 1000e^{kt}$, where k is constant. Given that when t is 0, the rate of change of the population is 100 per hour, find k.

16 A curve C is given by the equations $x = 2\cos t + \sin 2t$, $y = \cos t - 2\sin 2t$, $0 \leqslant t < \pi$, where t is a parameter.

(a) Find $\frac{dx}{dt}$ and $\frac{dy}{dt}$ in terms of t.

(b) Find the value of $\frac{dy}{dx}$ at the point P on C where $t = \frac{\pi}{4}$.

(c) Find an equation of the normal to the curve at P.

17 A curve has parametric equations $x = t^2 - 1$, $y = 2t + 2$, $t \in \mathbb{R}$. The normal to this curve at the point P given by $t = 2$ meets the curve again at Q. Find the value of t at Q. **E**

Vectors

5

What you should know

1 A vector is a quantity that has both magnitude and direction.

2 Vectors that are equal have both the same magnitude and the same direction.

3 Two vectors are added using the 'triangle law'.

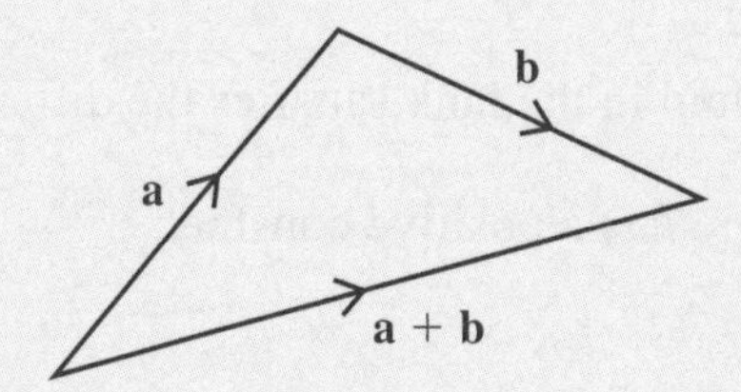

4 Adding the vectors $\overrightarrow{PQ}$ and $\overrightarrow{QP}$ gives the zero vector $\mathbf{0}$.
($\overrightarrow{PQ} + \overrightarrow{QP} = \mathbf{0}$)

5 The modulus of a vector is another name for its magnitude.
The modulus of the vector $\mathbf{a}$ is written as $|\mathbf{a}|$.
The modulus of the vector $\overrightarrow{PQ}$ is written as $|\overrightarrow{PQ}|$.

6 The vector $-\mathbf{a}$ has the same magnitude as the vector $\mathbf{a}$ but is in the opposite direction.

7 Any vector parallel to the vector $\mathbf{a}$ may be written as $\lambda\mathbf{a}$, where λ is a non-zero scalar.

8 $\mathbf{a} - \mathbf{b}$ is defined to be $\mathbf{a} + (-\mathbf{b})$.

9 A unit vector is a vector which has magnitude (or modulus) 1 unit.

10 If $\lambda\mathbf{a} + \mu\mathbf{b} = \alpha\mathbf{a} + \beta\mathbf{b}$, and the non-zero vectors $\mathbf{a}$ and $\mathbf{b}$ are not parallel, then $\lambda = \alpha$ and $\mu = \beta$.

11 The position vector of a point A is the vector $\overrightarrow{OA}$, where O is the origin. $\overrightarrow{OA}$ is usually written as vector $\mathbf{a}$.

12 $\overrightarrow{AB} = \mathbf{b} - \mathbf{a}$, where $\mathbf{a}$ and $\mathbf{b}$ are the position vectors of A and B respectively.

13 The vectors **i**, **j** and **k** are unit vectors parallel to the x-axis, the y-axis and the z-axis and in the direction of x increasing, y increasing and z increasing, respectively.

14 The modulus (or magnitude) of $x\mathbf{i} + y\mathbf{j}$ is $\sqrt{x^2 + y^2}$.

15 The vector $x\mathbf{i} + y\mathbf{j} + z\mathbf{k}$ may be written as a column matrix $\begin{pmatrix} x \\ y \\ z \end{pmatrix}$.

16 The distance from the origin to the point (x, y, z) is $\sqrt{x^2 + y^2 + z^2}$.

17 The distance between the points (x_1, y_1, z_1) and (x_2, y_2, z_2) is $\sqrt{(x_1 - x_2)^2 + (y_1 - y_2)^2 + (z_1 - z_2)^2}$.

18 The modulus (or magnitude) of $x\mathbf{i} + y\mathbf{j} + z\mathbf{k}$ is $\sqrt{x^2 + y^2 + z^2}$.

19 The scalar product of two vectors **a** and **b** is written as **a.b**, and defined by $\mathbf{a}.\mathbf{b} = |\mathbf{a}|\,|\mathbf{b}| \cos\theta$, where θ is the angle between **a** and **b**.

20 If **a** and **b** are the position vectors of the points A and B, then

$$\cos AOB = \frac{\mathbf{a}.\mathbf{b}}{|\mathbf{a}|\,|\mathbf{b}|}.$$

21 The non-zero vectors **a** and **b** are perpendicular if and only if $\mathbf{a}.\mathbf{b} = 0$.

22 If **a** and **b** are parallel, $\mathbf{a}.\mathbf{b} = |\mathbf{a}|\,|\mathbf{b}|$.
In particular, $\mathbf{a}.\mathbf{a} = |\mathbf{a}|^2$.

23 If $\mathbf{a} = a_1\mathbf{i} + a_2\mathbf{j} + a_3\mathbf{k}$ and $\mathbf{b} = b_1\mathbf{i} + b_2\mathbf{j} + b_3\mathbf{k}$,

$$\mathbf{a}.\mathbf{b} = \begin{pmatrix} a_1 \\ a_2 \\ a_3 \end{pmatrix} . \begin{pmatrix} b_1 \\ b_2 \\ b_3 \end{pmatrix} = a_1b_1 + a_2b_2 + a_3b_3.$$

24 A vector equation of a straight line passing through the point A with position vector **a**, and parallel to the vector **b**, is $\mathbf{r} = \mathbf{a} + t\mathbf{b}$, where t is a scalar parameter.

25 A vector equation of a straight line passing through the points C and D, with position vectors **c** and **d** respectively, is $\mathbf{r} = \mathbf{c} + t(\mathbf{d} - \mathbf{c})$, where t is a scalar parameter.

26 The acute angle θ between two straight lines is given by

$$\cos\theta = \left|\frac{\mathbf{a}.\mathbf{b}}{|\mathbf{a}|\,|\mathbf{b}|}\right|,$$ where **a** and **b** are direction vectors of the lines.

Test yourself

What to review

If your answer is incorrect

1 $\mathbf{a} = 2\mathbf{i} + 3\mathbf{j}$ and $\mathbf{b} = 3\mathbf{i} + \lambda\mathbf{j}$. Find the value of λ, given that:

(a) $\mathbf{a}$ and $\mathbf{b}$ are parallel,

(b) $\mathbf{a}$ and $\mathbf{b}$ are perpendicular.

Review Edexcel Book C4 pages 56 and 69–72
Revise for C4 page 44 Worked examination style question 1

---

2 Find a unit vector in the same direction as $2\mathbf{i} - 14\mathbf{j} + 5\mathbf{k}$.

Review Edexcel Book C4 pages 57, 62 and 67
Revise for C4 page 42 Example 4

---

3 $\mathbf{a} = 6\mathbf{i} + 5\mathbf{j} - \mathbf{k}$ and $\mathbf{b} = -2\mathbf{i} + 3\mathbf{j} + 5\mathbf{k}$.

(a) Find the value of $\mathbf{a}.\mathbf{b}$.

(b) Calculate the angle between $\mathbf{a}$ and $\mathbf{b}$, giving your answer in degrees to one decimal place.

Review Edexcel Book C4 pages 69–72
Revise for C4 page 44 Worked examination style question 1

---

4 For the parallelogram $ABCD$, the position vectors of A, B and C, referred to an origin O, are $\mathbf{a}$, $\mathbf{b}$ and $\mathbf{c}$ respectively.
Find, in terms of $\mathbf{a}$, $\mathbf{b}$ and $\mathbf{c}$ the position vectors of:

(a) D

(b) the mid-point of CD.

Review Edexcel Book C4 pages 53–55 and 60
Revise for C4 page 42 Example 3

---

5 The line l passes through the points with coordinates $(2, 7, -1)$ and $(8, -1, -5)$.

(a) Find a vector equation for l.

The point $(c, d, 0)$ also lies on l.

(b) Find the value of c and the value of d.

Review Edexcel Book C4 pages 75–78
Revise for C4 page 44 Worked examination style question 2

---

6 The lines l_1 and l_2 have vector equations:

$$l_1: \quad \mathbf{r} = \begin{pmatrix} 7 \\ 2 \\ -1 \end{pmatrix} + t\begin{pmatrix} 3 \\ 4 \\ 5 \end{pmatrix},$$

$$l_2: \quad \mathbf{r} = \begin{pmatrix} 7 \\ 2 \\ -1 \end{pmatrix} + s\begin{pmatrix} 5 \\ -4 \\ 3 \end{pmatrix}.$$

(a) Find the cosine of the acute angle between l_1 and l_2.

The point A lies on l_1 and the point B lies on l_2.
The y-coordinate of A is 6 and the y-coordinate of B is also 6.

(b) Find the exact distance from A to B.

Review Edexcel Book C4 pages 64–66 and 81
Revise for C4 page 43 Example 5 and page 46 Worked examination style question 3

Example 1

The vector **a** is directed due north and $|\mathbf{a}| = 6$.
The vector **b** is directed due east and $|\mathbf{b}| = 4$.

(a) Find the value of $|\mathbf{a} + 2\mathbf{b}|$.

(b) Find the direction of $\mathbf{a} + 2\mathbf{b}$, as a bearing to the nearest degree.

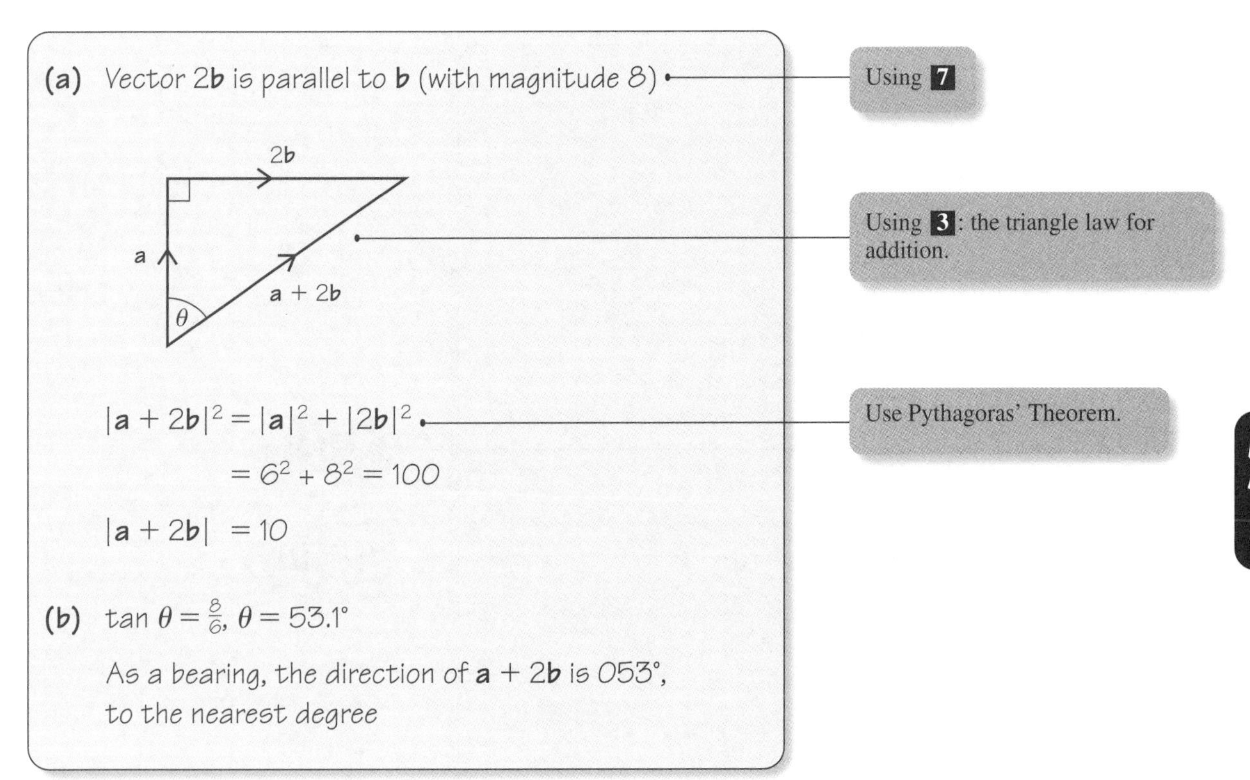

5

Example 2

Given that $5\lambda\mathbf{a} + (1 - \lambda)\mathbf{b} = \mu\mathbf{a} + (\mu - 3)\mathbf{b}$, where **a** and **b** are non-parallel, non-zero vectors, find the value of λ and the value of μ.

$5\lambda = \mu$ and $(1 - \lambda) = (\mu - 3)$ — Using **10**

So
$$1 - \lambda = 5\lambda - 3$$
$$6\lambda = 4$$
$$\lambda = \tfrac{2}{3}$$
$$\mu = \tfrac{10}{3}$$

Example 3

For the quadrilateral $OABC$, the position vectors of A, B and C, referred to the origin O, are $\mathbf{a}$, $\mathbf{b}$ and $\mathbf{c}$ respectively.
The point M is the mid-point of AB and the point N divides BC in the ratio 2 : 1.

Find an expression for $\overrightarrow{MN}$ in terms of $\mathbf{a}$, $\mathbf{b}$ and $\mathbf{c}$.

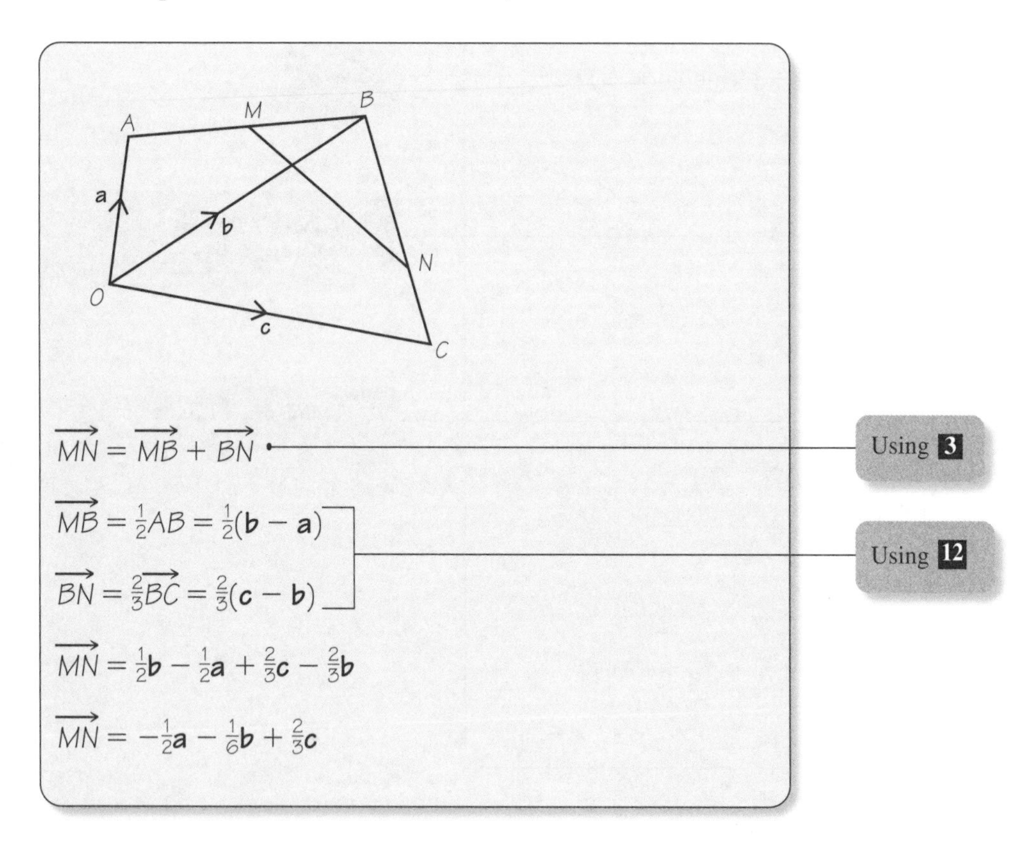

$\overrightarrow{MN} = \overrightarrow{MB} + \overrightarrow{BN}$ — Using **3**

$\overrightarrow{MB} = \frac{1}{2}AB = \frac{1}{2}(\mathbf{b} - \mathbf{a})$

$\overrightarrow{BN} = \frac{2}{3}\overrightarrow{BC} = \frac{2}{3}(\mathbf{c} - \mathbf{b})$ — Using **12**

$\overrightarrow{MN} = \frac{1}{2}\mathbf{b} - \frac{1}{2}\mathbf{a} + \frac{2}{3}\mathbf{c} - \frac{2}{3}\mathbf{b}$

$\overrightarrow{MN} = -\frac{1}{2}\mathbf{a} - \frac{1}{6}\mathbf{b} + \frac{2}{3}\mathbf{c}$

Example 4

Given that $\mathbf{a} = 3\mathbf{i} + 2\mathbf{j}$, $\mathbf{b} = 10\mathbf{i} + 12\mathbf{j}$ and $\mathbf{c} = -6\mathbf{i} + 10\mathbf{j}$, find $|\mathbf{a} + \mathbf{b} + \mathbf{c}|$, and hence find a unit vector in the same direction as $\mathbf{a} + \mathbf{b} + \mathbf{c}$.

$$\mathbf{a} + \mathbf{b} + \mathbf{c} = 3\mathbf{i} + 2\mathbf{j} + 10\mathbf{i} + 12\mathbf{j} - 6\mathbf{i} + 10\mathbf{j}$$

$$= 7\mathbf{i} + 24\mathbf{j}$$

$|\mathbf{a} + \mathbf{b} + \mathbf{c}| = \sqrt{(7^2 + 24^2)} = 25$ — Using **14**

$\text{Unit vector} = \dfrac{\mathbf{a} + \mathbf{b} + \mathbf{c}}{|\mathbf{a} + \mathbf{b} + \mathbf{c}|} = \frac{1}{25}(7\mathbf{i} + 24\mathbf{j})$ — Using **9**

Example 5

Calculate the distance between the points (3, 6, −4) and (7, 1, 3), giving your answer in its simplest form.

$$\text{Distance} = \sqrt{(3-7)^2 + (6-1)^2 + (-4-3)^2}$$
$$= \sqrt{(16 + 25 + 49)} = \sqrt{90}$$
$$= \sqrt{9}\sqrt{10} = 3\sqrt{10}$$

Using **17**

Example 6

The points A and B have position vectors $(5\mathbf{i} - 2\mathbf{j} + 3\mathbf{k})$ and $(8\mathbf{i} + 8\mathbf{j} + t\mathbf{k})$ respectively, and $|\overrightarrow{AB}| = 5\sqrt{5}$.
Find the possible values of t.

$$\overrightarrow{AB} = \begin{pmatrix} 8 \\ 8 \\ t \end{pmatrix} - \begin{pmatrix} 5 \\ -2 \\ 3 \end{pmatrix} = \begin{pmatrix} 3 \\ 10 \\ t-3 \end{pmatrix}$$

Using **12** and **15**

$$|\overrightarrow{AB}| = \sqrt{3^2 + 10^2 + (t-3)^2}$$

Using **18**

$$= \sqrt{(9 + 100 + t^2 - 6t + 9)}$$
$$= \sqrt{(t^2 - 6t + 118)}$$

So $\sqrt{(t^2 - 6t + 118)} = 5\sqrt{5}$

$$t^2 - 6t + 118 = 125$$

Square both sides.

$$t^2 - 6t - 7 = 0$$
$$(t-7)(t+1) = 0$$
$$t = 7 \text{ or } t = -1$$

Worked examination style question 1

Given that $\mathbf{a} = 7\mathbf{i} - 4\mathbf{j} + 2\mathbf{k}$ and $\mathbf{b} = 2\mathbf{i} + 8\mathbf{j} + t\mathbf{k}$:

(a) find the value of t for which **a** and **b** are perpendicular,

(b) find, to the nearest degree, the angle between **a** and **b** when $t = 11$.

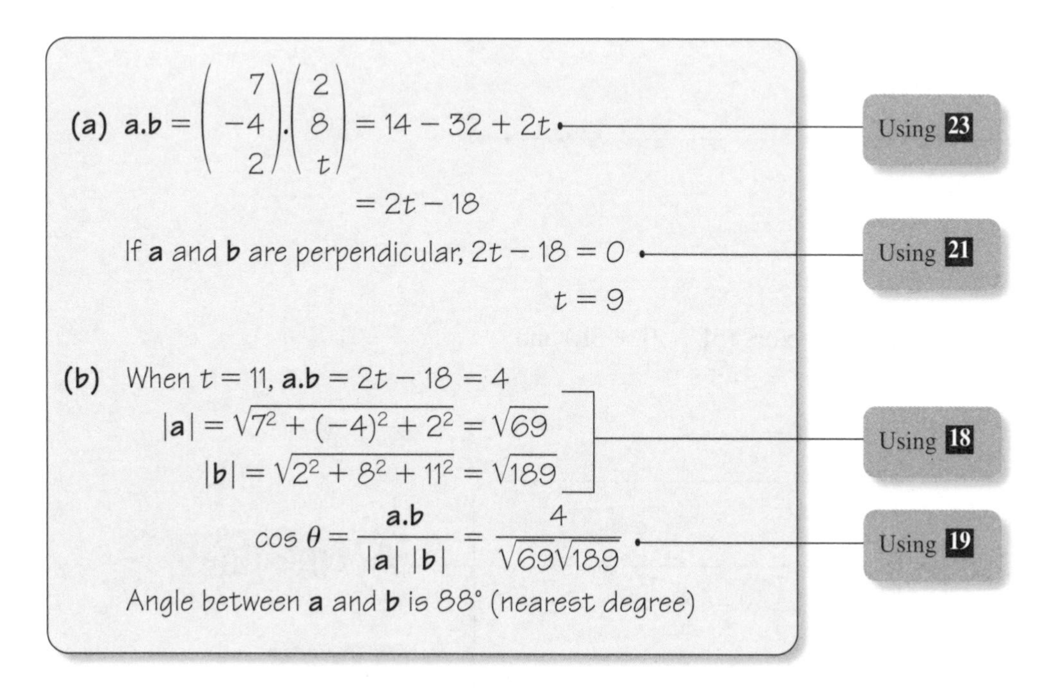

Worked examination style question 2

The points A and B have position vectors $(5\mathbf{i} + 8\mathbf{j} - 4\mathbf{k})$ and $(8\mathbf{i} + 2\mathbf{j} + 5\mathbf{k})$ respectively.

(a) Find a vector equation for the line l which passes through A and B.

(b) Given that the point with coordinates $(p, 4p, q)$ lies on l, find the value of p and the value of q.

(a) $\mathbf{a} = \begin{pmatrix} 5 \\ 8 \\ -4 \end{pmatrix}$, $\mathbf{b} = \begin{pmatrix} 8 \\ 2 \\ 5 \end{pmatrix}$

$$\mathbf{b} - \mathbf{a} = \begin{pmatrix} 8 \\ 2 \\ 5 \end{pmatrix} - \begin{pmatrix} 5 \\ 8 \\ -4 \end{pmatrix} = \begin{pmatrix} 3 \\ -6 \\ 9 \end{pmatrix}$$

Find a direction vector for the line.

Equation of ℓ: $\mathbf{r} = \begin{pmatrix} 5 \\ 8 \\ -4 \end{pmatrix} + t\begin{pmatrix} 3 \\ -6 \\ 9 \end{pmatrix}$

Using 25

$\left(\text{Alternative: } \mathbf{r} = \begin{pmatrix} 5 \\ 8 \\ -4 \end{pmatrix} + t\begin{pmatrix} 1 \\ -2 \\ 3 \end{pmatrix}\right)$

$\begin{pmatrix} 3 \\ -6 \\ 9 \end{pmatrix} = 3\begin{pmatrix} 1 \\ -2 \\ 3 \end{pmatrix}$, so the vectors are parallel.

(b) $\begin{pmatrix} 5 \\ 8 \\ -4 \end{pmatrix} + t\begin{pmatrix} 1 \\ -2 \\ 3 \end{pmatrix} = \begin{pmatrix} p \\ 4p \\ q \end{pmatrix}$

So $p = 5 + t$ ①

$4p = 8 - 2t$ ②

$q = -4 + 3t$ ③

From ①, $t = p - 5$

Substituting into ②:

$4p = 8 - 2(p - 5)$

$4p = 8 - 2p + 10$

$6p = 18$

$p = 3$

So $t = p - 5 = -2$

and $q = -4 + 3t = -10$

$p = 3, q = -10$

Worked examination style question 3

The lines l_1 and l_2 have vector equations

$$\text{and } \mathbf{r} = \begin{pmatrix} 5 \\ 0 \\ 4 \end{pmatrix} + t\begin{pmatrix} 3 \\ -4 \\ 2 \end{pmatrix}$$

$$\text{and } \mathbf{r} = \begin{pmatrix} 5 \\ -1 \\ 9 \end{pmatrix} + s\begin{pmatrix} 2 \\ -3 \\ 3 \end{pmatrix} \text{ respectively.}$$

(a) Show that l_1 and l_2 intersect.

(b) Find the coordinates of their point of intersection.

(c) Find the acute angle between l_1 and l_2, giving your answer in degrees to 1 decimal place.

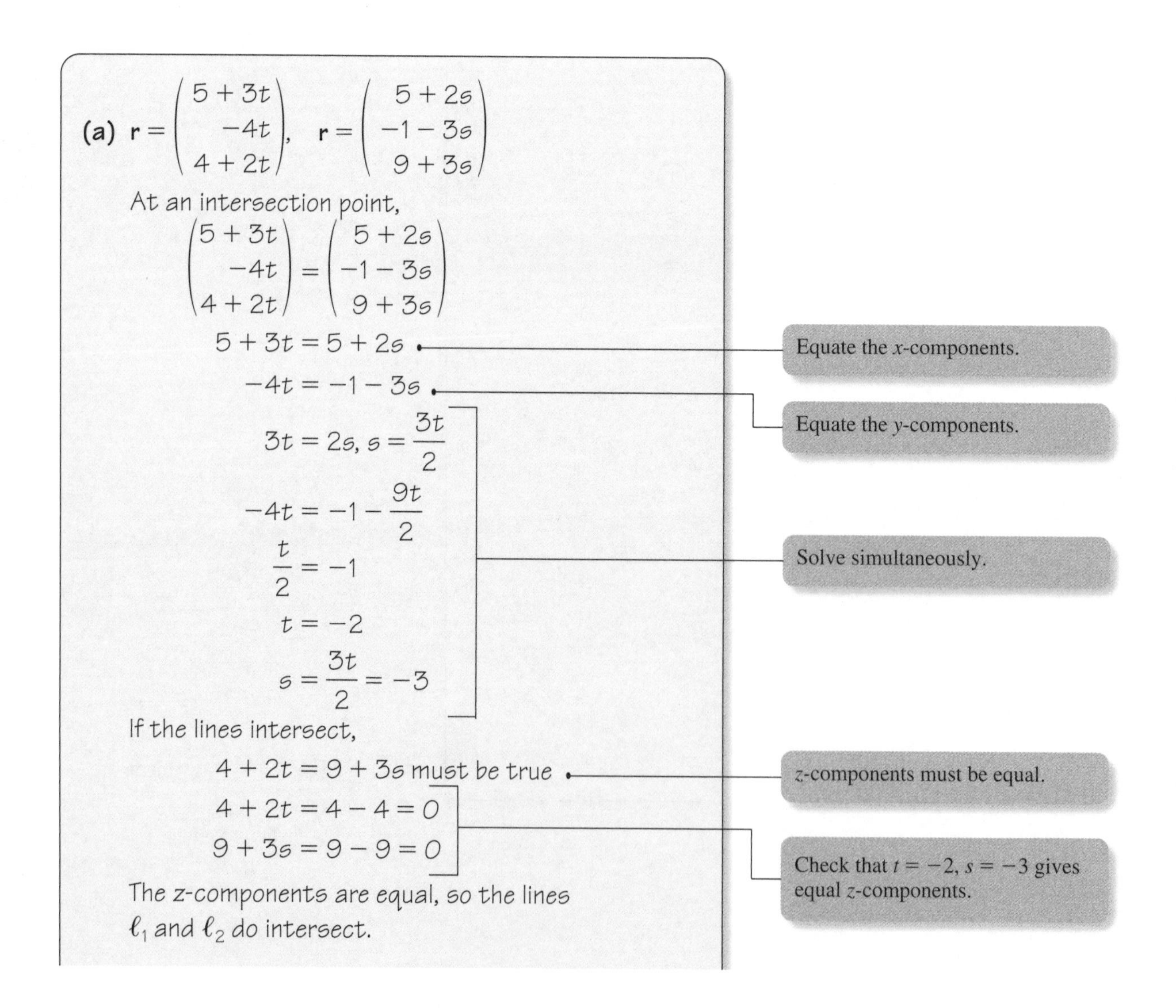

(a) $\mathbf{r} = \begin{pmatrix} 5+3t \\ -4t \\ 4+2t \end{pmatrix}$, $\mathbf{r} = \begin{pmatrix} 5+2s \\ -1-3s \\ 9+3s \end{pmatrix}$

At an intersection point,

$$\begin{pmatrix} 5+3t \\ -4t \\ 4+2t \end{pmatrix} = \begin{pmatrix} 5+2s \\ -1-3s \\ 9+3s \end{pmatrix}$$

$5 + 3t = 5 + 2s$ — Equate the x-components.

$-4t = -1 - 3s$ — Equate the y-components.

$3t = 2s$, $s = \frac{3t}{2}$

$-4t = -1 - \frac{9t}{2}$

$\frac{t}{2} = -1$

$t = -2$

$s = \frac{3t}{2} = -3$

Solve simultaneously.

If the lines intersect,

$4 + 2t = 9 + 3s$ must be true — z-components must be equal.

$4 + 2t = 4 - 4 = 0$

$9 + 3s = 9 - 9 = 0$

Check that $t = -2$, $s = -3$ gives equal z-components.

The z-components are equal, so the lines ℓ_1 and ℓ_2 do intersect.

(b) Intersection point: $\mathbf{r} = \begin{pmatrix} 5+3t \\ -4t \\ 4+2t \end{pmatrix} = \begin{pmatrix} -1 \\ 8 \\ 0 \end{pmatrix}$

Coordinates: $(-1, 8, 0)$

(c) $\mathbf{a} = \begin{pmatrix} 3 \\ -4 \\ 2 \end{pmatrix}$, $\mathbf{b} = \begin{pmatrix} 2 \\ -3 \\ 3 \end{pmatrix}$ — Use the direction vectors.

$\mathbf{a.b} = \begin{pmatrix} 3 \\ -4 \\ 2 \end{pmatrix} . \begin{pmatrix} 2 \\ -3 \\ 3 \end{pmatrix} = 6 + 12 + 6 = 24$ — Using **23**

$|\mathbf{a}| = \sqrt{3^2 + (-4)^2 + 2^2} = \sqrt{29}$

$|\mathbf{b}| = \sqrt{2^2 + (-3)^2 + 3^2} = \sqrt{22}$ — Using **18**

$\cos\theta = \dfrac{\mathbf{a.b}}{|\mathbf{a}|\,|\mathbf{b}|} = \dfrac{24}{\sqrt{29}\sqrt{22}}$ — Using **26**

Angle between ℓ_1 and ℓ_2 is 18.2° (1 d.p.)

5

Revision exercise 5

1 The diagram shows a pentagon *PQRST* in which *SR* is parallel to *PQ* and $SR = 2PQ$. The point *M* is the mid-point of *PT*. $\overrightarrow{PQ} = \mathbf{a}$, $\overrightarrow{QR} = \mathbf{b}$ and $\overrightarrow{ST} = \mathbf{c}$. Find, in terms of **a**, **b** and **c**:

(a) $\overrightarrow{PT}$,

(b) $\overrightarrow{RM}$.

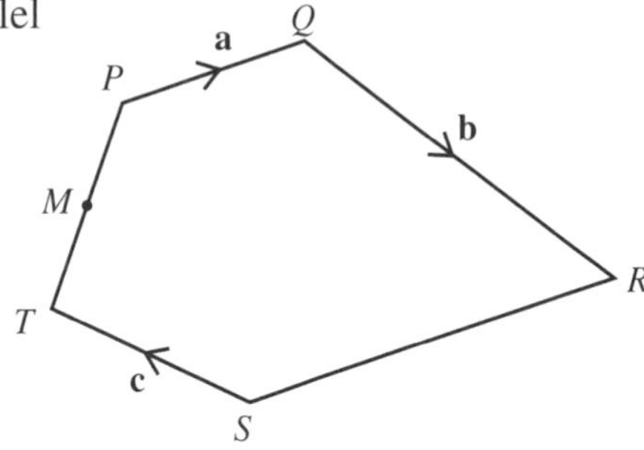

2 Given that $\mathbf{a} = 6\mathbf{i} - 5\mathbf{j}$, $\mathbf{b} = 8\mathbf{i} - 6\mathbf{j}$ and $\mathbf{c} = 10\mathbf{i} - 7\mathbf{j}$:
(a) show that $\mathbf{a} + \mathbf{b} + \mathbf{c}$ is parallel to **b**,
(b) find $|\mathbf{a} + \mathbf{b} + \mathbf{c}|$.

3 Find, to 3 significant figures, the distance from the origin to the point with coordinates (5, 13, −2).

4 The points *A* and *B* have position vectors $(\lambda\mathbf{i} + 2\mathbf{j} + 5\mathbf{k})$ and $(3\mathbf{i} + 7\mathbf{j} + 11\mathbf{k})$ respectively, referred to the origin *O*. Given that $\angle OAB = 90°$, calculate the possible values of λ.

5 Given that **a** and **b** are perpendicular vectors, and that $|\mathbf{a}| = 3$ and $|\mathbf{b}| = 5$, find the value of $(\mathbf{a} + \mathbf{b}).(\mathbf{a} + \mathbf{b})$.

6 Find, to the nearest tenth of a degree, the acute angle between the x-axis and the line with vector equation

$$\mathbf{r} = 2\mathbf{i} + \lambda(5\mathbf{i} + 2\mathbf{j} - \mathbf{k}).$$

7 The points A and B have position vectors $(3\mathbf{i} + t\mathbf{j} + 5\mathbf{k})$ and $(7\mathbf{i} + \mathbf{j} + t\mathbf{k})$ respectively.

(a) Find $|\overrightarrow{AB}|$ in terms of t.

(b) Find the value of t that makes $|\overrightarrow{AB}|$ a minimum.

(c) Find the minimum value of $|\overrightarrow{AB}|$.

8 In the diagram, $\overrightarrow{OA} = \mathbf{a}$, $\overrightarrow{OB} = \mathbf{b}$, $\overrightarrow{OP} = 3\mathbf{a}$ and $\overrightarrow{OQ} = 2\mathbf{b}$.

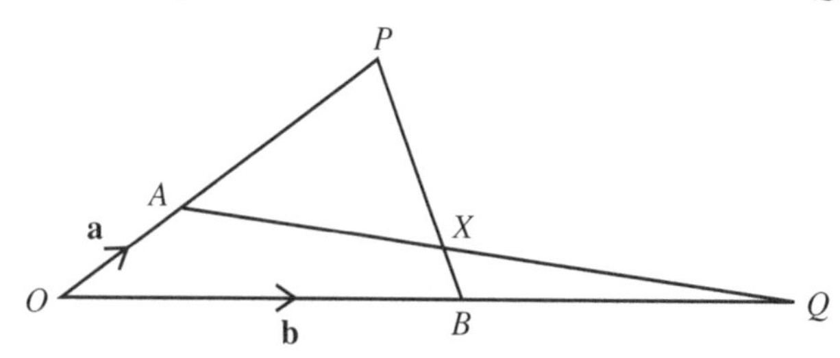

(a) Find $\overrightarrow{BP}$ and $\overrightarrow{AQ}$ in terms of $\mathbf{a}$ and $\mathbf{b}$.

Given that $\overrightarrow{BX} = \lambda\overrightarrow{BP}$ and $\overrightarrow{AX} = \mu\overrightarrow{AQ}$, where X is the intersection point of AQ and BP,

(b) find two expressions for $\overrightarrow{OX}$, one in terms of $\mathbf{a}$, $\mathbf{b}$ and λ and the other in terms of $\mathbf{a}$, $\mathbf{b}$ and μ.

(c) Hence find the ratio $BX : XP$ and the ratio $AX : XQ$.

9 The line l_1 passes through the point with coordinates $(1, 2, 5)$ and is parallel to the vector $(-3\mathbf{i} + 2\mathbf{j} + 4\mathbf{k})$.
The line l_2 passes through the points $(3, 4, -1)$ and $(6, -8, 5)$.

(a) Find vector equations for l_1 and l_2.

(b) Show that l_1 and l_2 intersect, and find the coordinates of their point of intersection.

10 The equations of the lines l_1 and l_2 are given by

$$l_1: \quad \mathbf{r} = \mathbf{i} + 3\mathbf{j} + 5\mathbf{k} + \lambda(\mathbf{i} + 2\mathbf{j} - \mathbf{k}),$$
$$l_2: \quad \mathbf{r} = -2\mathbf{i} + 3\mathbf{j} - 4\mathbf{k} + \mu(2\mathbf{i} + \mathbf{j} + 4\mathbf{k}),$$

where λ and μ are parameters.

(a) Show that l_1 and l_2 intersect and find the coordinates of Q, their point of intersection.

(b) Show that l_1 is perpendicular to l_2.

The point P with x-coordinate 3 lies on the line l_1 and the point R with x-coordinate 4 lies on the line l_2.

(c) Find, in its simplest form, the exact area of the triangle PQR. **E**

11 Relative to a fixed origin O, the point A has position vector $3\mathbf{i} + 2\mathbf{j} - \mathbf{k}$, the point B has position vector $5\mathbf{i} + \mathbf{j} + \mathbf{k}$, and the point C has position vector $7\mathbf{i} - \mathbf{j}$.

(a) Find the cosine of angle ABC.

(b) Find the exact value of the area of triangle ABC.

The point D has position vector $7\mathbf{i} + 3\mathbf{k}$.

(c) Show that AC is perpendicular to CD.

(d) Find the ratio $AD : DB$. 

12 Relative to a fixed origin O, the point A has position vector $4\mathbf{i} + 8\mathbf{j} - \mathbf{k}$, and the point B has position vector $7\mathbf{i} + 14\mathbf{j} + 5\mathbf{k}$.

(a) Find the vector $\overrightarrow{AB}$.

(b) Calculate the cosine of $\angle OAB$.

(c) Show that, for all values of λ, the point P with position vector $\lambda\mathbf{i} + 2\lambda\mathbf{j} + (2\lambda - 9)\mathbf{k}$ lies on the line through A and B.

(d) Find the value of λ for which OP is perpendicular to AB.

(e) Hence find the coordinates of the foot of the perpendicular from O to AB. 

Integration 6

What you should know

1 You should be familiar with the following integrals. (The $+c$ has been omitted here but should be included in the examination.)

	$\mathbf{f(x)}$	$\int \mathbf{f(x)\,dx}$
(i)	x^n	$\dfrac{x^{n+1}}{n+1}$
(ii)	e^x	e^x
(iii)	$\dfrac{1}{x}$	$\ln\lvert x\rvert$
(iv)	$\sin x$	$-\cos x$
(v)	$\cos x$	$\sin x$
(vi)	$\tan x$	$\ln\lvert\sec x\rvert$
(vii)	$\cot x$	$\ln\lvert\sin x\rvert$
(viii)	$\sec x$	$\ln\lvert\sec x + \tan x\rvert$
(ix)	$\operatorname{cosec} x$	$-\ln\lvert\operatorname{cosec} x + \cot x\rvert$
(x)	$\sec^2 x$	$\tan x$
(xi)	$-\operatorname{cosec} x \cot x$	$\operatorname{cosec} x$
(xii)	$\sec x \tan x$	$\sec x$
(xiii)	$-\operatorname{cosec}^2 x$	$\cot x$

2 Using the chain rule in reverse you can obtain generalizations of the above formulae.

(i) $\displaystyle\int f'(ax+b)\,dx = \frac{1}{a}f(ax+b) + c$

(ii) $\displaystyle\int (ax+b)^n\,dx = \frac{1}{a}\frac{(ax+b)^{n+1}}{n+1} + c$

(iii) $\displaystyle\int e^{ax+b}\,dx = \frac{1}{a}e^{ax+b} + c$

(iv) $\displaystyle\int \frac{1}{ax+b}\,dx = \frac{1}{a}\ln|ax+b| + c$

(v) $\displaystyle\int \cos(ax+b)\,dx = \frac{1}{a}\sin(ax+b) + c$

(vi) $\displaystyle\int \sin(ax+b)\,dx = -\frac{1}{a}\cos(ax+b) + c$

(vii) $\displaystyle\int \sec^2(ax+b)\,dx = \frac{1}{a}\tan(ax+b) + c$

(viii) $\displaystyle\int \operatorname{cosec}(ax+b)\cot(ax+b)\,dx = -\frac{1}{a}\operatorname{cosec}(ax+b) + c$

(ix) $\displaystyle\int \operatorname{cosec}^2(ax+b)\,dx = -\frac{1}{a}\cot(ax+b) + c$

(x) $\displaystyle\int \sec(ax+b)\tan(ax+b)\,dx = \frac{1}{a}\sec(ax+b) + c$

3 Sometimes trigonometric identities can be useful to help change the expression into one you know how to integrate. e.g. to integrate $\sin^2 x$ or $\cos^2 x$ use formulae for $\cos 2x$, so

$$\int \sin^2 x\,dx = \int (\tfrac{1}{2} - \tfrac{1}{2}\cos 2x)\,dx.$$

4 You can use partial fractions to integrate expressions of the type $\dfrac{x+8}{(x-1)(x+2)}$.

5 The following general patterns should be remembered.

$$\int \frac{f'(x)}{f(x)}\,dx = \ln|f(x)| + c$$

$$\int f'(x)[f(x)]^n\,dx = \frac{1}{n+1}[f(x)]^{n+1};\ n \neq -1$$

6 Sometimes you can simplify an integral by changing the variable. This process is called **integration by substitution.**

7 **Integration by parts** is done as follows

$$\int u\frac{dv}{dx}\,dx = uv - \int v\frac{du}{dx}\,dx.$$

8 The area of region enclosed by $y = f(x)$, the x-axis and $x = a$ and $x = b$ is given by

$$\text{area} = \int_a^b y\,dx.$$

9 The volume of revolution about the x-axis between $x = a$ and $x = b$ is given by

$$\text{volume} = \pi\int_a^b y^2\,dx.$$

10 Separation of variables for first order differential equations

$$\frac{dy}{dx} = f(x)\,g(y) \Rightarrow \int \frac{1}{g(y)}\,dy = \int f(x)\,dx.$$

11 The following integrals are given in the formula booklet but you need to be able to identify them.

$$\int \tan x\,dx = \ln|\sec x| + c$$

$$\int \sec x\,dx = \ln|\sec x + \tan x| + c$$

$$\int \cot x\,dx = \ln|\sin x| + c$$

$$\int \operatorname{cosec} x\,dx = -\ln|\operatorname{cosec} x + \cot x| + c$$

12 You may have to use the trapezium rule (as in C2 but applied to more complicated functions).
The **trapezium rule** (in the formula booklet) is

$$\int_a^b y\,dx \approx \tfrac{1}{2}h[y_0 + 2(y_1 + y_2 + \ldots + y_{n-1}) + y_n]$$

where $h = \dfrac{b-a}{n}$ and $y_i = f(a + ih)$.

Test yourself	What to review
	If your answer is incorrect
1 Find $\int \left(2\cos(3x-1) + \dfrac{1}{3x-1}\right) dx$.	*Review Edexcel Book C4 pages 88–91* *Revise for C4 page 53 Example 1*
2 Find $\int (1 + \tan x)^2\,dx$.	*Review Edexcel Book C4 pages 92–94, 108* *Revise for C4 page 54 Example 3*
3 Find $\int \dfrac{x^3}{(x-1)(x+1)}\,dx$.	*Review Edexcel Book C4 pages 95–97* *Revise for C4 page 59 Worked examination style question 6*

4 Use the substitution $x = \tan u$ to show that

$$\int_0^{\sqrt{3}} \frac{1}{\sqrt{1+x^2}}\,dx = \ln|2+\sqrt{3}|.$$

Review Edexcel Book C4 pages 98–104 and 108
Revise for C4 page 55 Example 4 and page 55 Worked examination style question 1

---

5 Find $\int xe^{3x}\,dx$.

Review Edexcel Book C4 pages 105–107
Revise for C4 page 56 Worked examination style question 3

---

6 Solve the differential equation $2\tan x\frac{dy}{dx} + y^2 = 1$ given that $y = \frac{1}{2}$ when $x = \frac{\pi}{6}$.

Give your answer in the form $y = f(x)$.

Review Edexcel Book C4 pages 114–117
Revise for C4 page 56 Worked examination style question 6

6

Example 1

Find: **(a)** $\int \sec^2(4x+1)\,dx$ **(b)** $\int \frac{3}{1+2x}\,dx$.

(a) Let $I = \int \sec^2(4x+1)\,dx$

then $I = \frac{1}{4}\tan(4x+1) + c$

Using **2** (vii)

(b) Let $I = \int \frac{3}{1+2x}\,dx$

Consider $y = \ln|1+2x|$

Compare $\int \frac{1}{x}\,dx = \ln|x| + c$.

then $\frac{dy}{dx} = \int \frac{1}{1+2x} \times 2$

Remember the $\times 2$ from the chain rule.

So $I = \frac{3}{2}\ln|1+2x| + c$

You want $\frac{3}{1+2x}$, so adjust the constant.

You should recognise the forms $\frac{f'(x)}{f(x)}$ and $[f(x)]^n f'(x)$.

Example 2

Find:

(a) $\int \left(\frac{x+1}{x^2+2x+5}\right) dx$

(b) $\int \cos(4x-1)\sin^2(4x^2-1)\,dx$.

(a) $I = \int \left(\frac{x+1}{x^2+2x+5}\right) dx$

Let $y = \ln|x^2+2x+5|$

$\frac{dy}{dx} = \frac{2x+2}{x^2+2x+5} = \frac{2(x+1)}{x^2+2x+5}$

so $I = \frac{1}{2}\ln|x^2+2x+5| + c$

Using **5**: $\int \left(\frac{f'(x)}{f(x)}\right) dx = \ln|f(x)| + c$ and adjust constant.

(b) $I = \int \cos(4x-1)\sin^2(4x-1)\,dx$

Let $y = \sin^3(4x-1)$

$\frac{dy}{dx} = 3\sin^2(4x-1)\cos(4x-1) \times 4$

so $I = \frac{1}{12}\sin^3(4x-1) + c$

Using **5**: $\int [f(x)]^n f'(x)\,dx = \frac{1}{n+1}[f(x)]^{n+1} + c$ and adjust constant.

Remember the $\times 4$ from the chain rule .

Sometimes you may have to use trigonometric identities from C3 before you can integrate.

Example 3

Find $\int \cot^2 3x\,dx$.

$\cot^2 3x = \text{cosec}^2 3x - 1$

so $I = \int \cot^2 3x\,dx = \int (\text{cosec}^2 3x - 1)\,dx$

so $I = -\frac{1}{3}\cot(3x-1) + c$

Use $\text{cosec}^2 A = \cot^2 A + 1$.

Using **2**(ix)

Sometimes you can use integration by substitution.

Example 4

Find letting $u = \tan x$, find $\int \sec^2 x\, e^{\tan x}\, dx$.

$$I = \int \sec^2 e^{\tan x}\, dx$$

Let $u = \tan x$

$$\frac{du}{dx} = \sec^2 x$$

so $\sec^2 x\, dx = du$

and $I = \int e^u\, du$

so $I = e^u + c$

$$I = e^{\tan x} + c$$

Find $\frac{du}{dx}$ and rearrange to replace the dx term. In this case you can replace $\sec^2 x\, dx$ with du.

Write I in terms of a function of $u\, du$.

Integrate and substitute back to obtain the answer in terms of x.

Worked examination style question 1

Use the substitution $u = 1 + \sin x$ and integration to show that

$$\int \sin x \cos x(1 + \sin x)^5\, dx = \frac{(1 + \sin x)^6}{42}(6 \sin x - 1) + c.$$

$$I = \int \sin x \cos x(1 + \sin x)^5\, dx$$

Let $u = 1 + \sin x$

$$\frac{du}{dx} = \cos x$$

So $\cos x\, dx = du$ and $\sin x = u - 1$

So $I = \int (u - 1)u^5\, du$

$$I = \int (u^6 - u^5)\, du$$

$$I = \left[\frac{u^7}{7} - \frac{u^6}{6} + c\right]$$

$$I = \frac{u^6}{42}(6u - 7) + c$$

$$I = \frac{(1 + \sin x)^6}{42}(6 \sin x + 6 - 7) + c$$

$$I = \frac{(1 + \sin x)^6}{42}(6 \sin x - 1) + c$$

Find $\frac{du}{dx}$ and rearrange for dx.

In this case you can substitute for $\cos x\, dx$.

Write I in terms of u.

Multiply out and integrate.

Factorise and then substitute back in terms of x.

Since the question is a 'show that' make sure that you explain each step of the working to obtain the given answer. This example shows you the sort of working required.

Sometimes you may be asked to use a u^2 substitution (see Revision Exercise 6 question 12) or in simple cases you may be expected to choose your own substitution.

6

Worked examination style question 2

Use integration by substitution to evaluate $\int_0^3 15x\sqrt{x+1}\,\mathrm{d}x$.

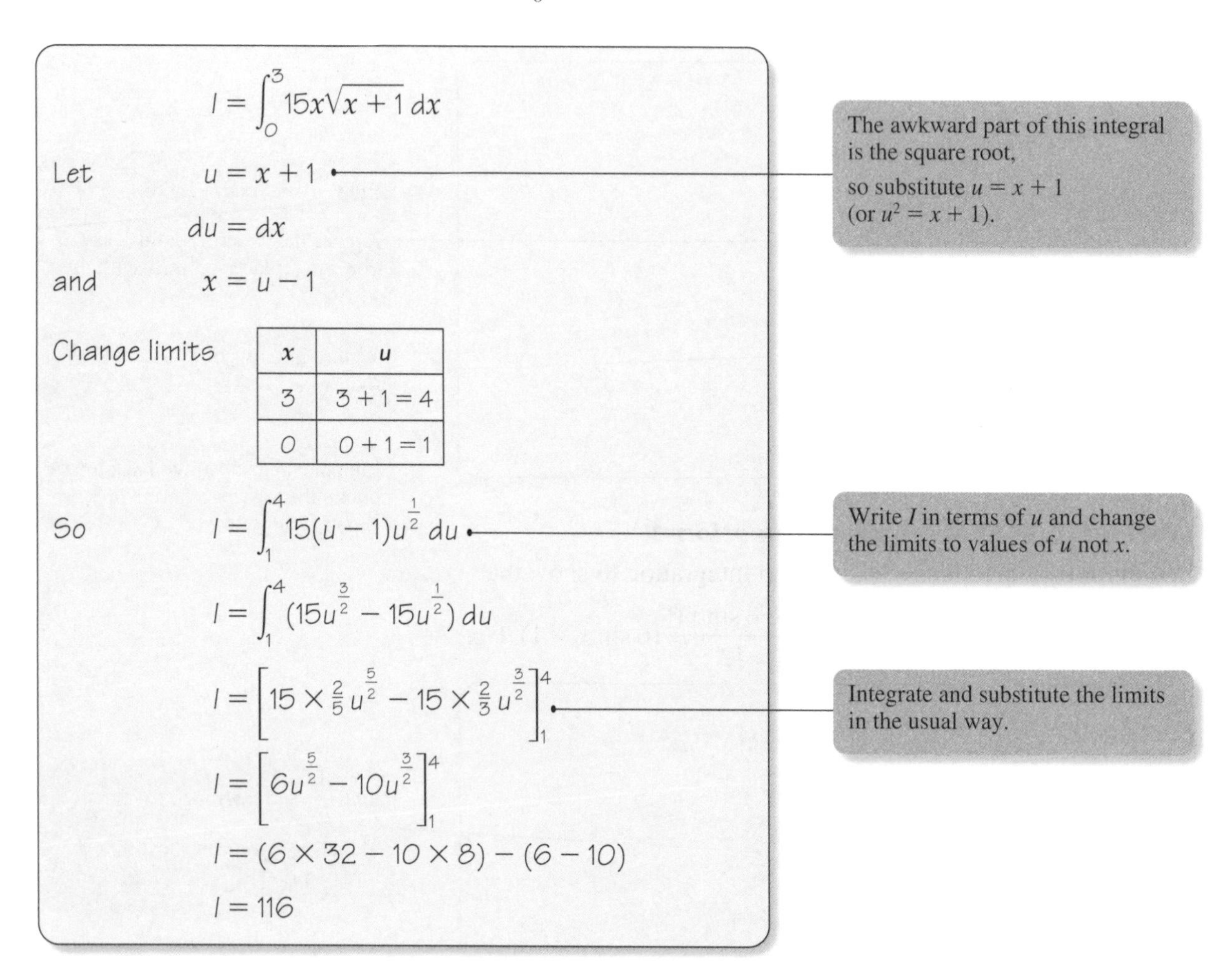

$I = \int_0^3 15x\sqrt{x+1}\,dx$

Let $u = x + 1$

$du = dx$

and $x = u - 1$

Change limits

x	u
3	$3 + 1 = 4$
0	$0 + 1 = 1$

So $I = \int_1^4 15(u-1)u^{\frac{1}{2}}\,du$

$I = \int_1^4 (15u^{\frac{3}{2}} - 15u^{\frac{1}{2}})\,du$

$I = \left[15 \times \frac{2}{5}u^{\frac{5}{2}} - 15 \times \frac{2}{3}u^{\frac{3}{2}}\right]_1^4$

$I = \left[6u^{\frac{5}{2}} - 10u^{\frac{3}{2}}\right]_1^4$

$I = (6 \times 32 - 10 \times 8) - (6 - 10)$

$I = 116$

The awkward part of this integral is the square root, so substitute $u = x + 1$ (or $u^2 = x + 1$).

Write I in terms of u and change the limits to values of u not x.

Integrate and substitute the limits in the usual way.

Worked examination style question 3

Use integration by parts to find the exact value of $\int_{\frac{\pi}{6}}^{\frac{\pi}{3}} x\,\mathrm{cosec}^2 x\,\mathrm{d}x$.

$I = \int_{\frac{\pi}{6}}^{\frac{\pi}{3}} x\,\mathrm{cosec}^2\,x\,dx$

$u = x \quad \Rightarrow \quad \frac{du}{dx} = 1$

$v = -\cot x \quad \Leftarrow \quad \frac{dv}{dx} = \mathrm{cosec}^2\,x$

$$I = \Big[-x\cot x\Big]_{\frac{\pi}{6}}^{\frac{\pi}{3}} - \int_{\frac{\pi}{6}}^{\frac{\pi}{3}} (-\cot x)\,dx$$

Using **7**: with $u = x$ and completing the usual table.

$$I = \left(-\frac{\pi}{3}\frac{1}{\sqrt{3}}\right) - \left(-\frac{\pi}{6}\sqrt{3}\right) + \Big[\ln|\sin x|\Big]_{\frac{\pi}{6}}^{\frac{\pi}{3}}$$

Use the 1, 2, $\sqrt{3}$ triangle to evaluate cot and sin.

$$I = \frac{\pi\sqrt{3}}{18} + \left(\ln\left(\frac{\sqrt{3}}{2}\right) - \ln\left(\frac{1}{2}\right)\right)$$

$$I = \frac{\pi\sqrt{3}}{18} + \ln\sqrt{3}$$

Sometimes you may have to use integration by parts twice (see Revision Exercise 6 question 18).

You can use the trapezium rule to evaluate integrals.

Worked examination style question 4

(a) Complete the table below, giving your answers to 3 significant figures.

x	1	1.5	2	2.5	3
$1 + \ln 2x$		2.10		2.61	2.79

(b) Use the trapezium rule, with four intervals, to estimate the value of

$$I = \int_1^3 (1 + \ln 2x)\,dx.$$

(c) Find the exact value of I.

(d) Calculate the percentage error in using your answer from part **(b)** to estimate the value of I.

(a)

x	1	1.5	2	2.5	3
$1 + \ln 2x$	**1.69**	2.10	**2.39**	2.61	2.79

(b) $I \approx \frac{1}{2} \times 0.5 \times [1.69 + 2.79 + 2(2.10 + 2.39 + 2.61)]$

So $I \approx \frac{1}{4}[18.68]$

$= 4.67$

(c) $I = \left[x\right]_1^3 + \int_1^3 \ln 2x \, dx$

Integrate each term separately.

$u = \ln 2x \Rightarrow \frac{du}{dx} = \frac{1}{x}$

$v = x \Leftarrow \frac{dv}{dx} = 1$

Integrate $\ln 2x$ by parts. Since it involves a ln function set $u = \ln 2x$ and $\frac{dv}{dx} = 1$.

$$\int_1^3 \ln 2x \, dx = \left[x \ln 2x\right]_1^3 - \int_1^3 x \frac{1}{x} \, dx$$

$$= (3 \ln 6) - (\ln 2) - \left[x\right]_1^3$$

So $I = 2 + \ln\left(\frac{6^3}{2}\right) - 2$

$I = \ln 108$

(d) Percentage error $= 100 \times \left(\frac{\ln 108 - 4.67}{\ln 108}\right) = 0.26\%$

You can use integration to find areas and volumes.

Worked examination style question 5

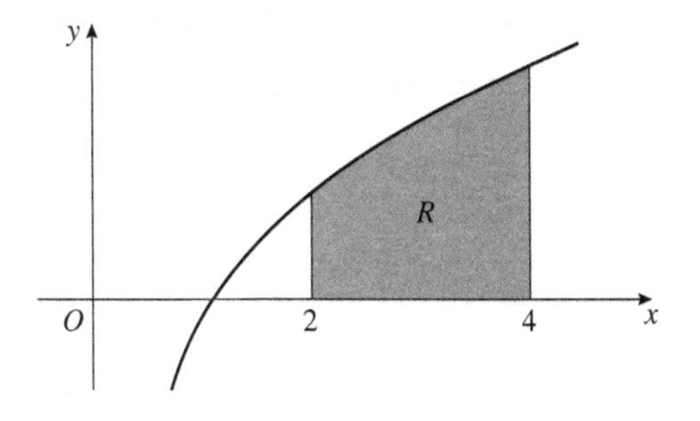

The figure shows part of the curve with equation $y = 4x - \frac{6}{x}, x > 0$.

The shaded region R is bounded by the curve, the x-axis and the lines with equations $x = 2$ and $x = 4$.
This region is rotated through 2π radians about the x-axis.
Find the exact value of the volume of the solid generated.

$$V = \pi \int_2^4 y^2 \, dx$$

$$y^2 = 16x^2 - 48 + 36x^{-2}$$

So $V = \pi \int_2^4 (16x^2 - 48 + 36x^{-2}) \, dx$

$$V = \pi \left[16 \frac{x^3}{3} - 48x - 36x^{-1}\right]_2^4$$

$$V = \pi \left[\left(\frac{1024}{3} - 192 - 9\right) - \left(\frac{128}{3} - 96 - 18\right)\right]$$

$$V = \pi \left[\frac{896}{3} - 87\right] = 211\tfrac{2}{3}\,\pi$$

You can solve differential equations by separation of variables.

Worked examination style question 6

Obtain the solution of $\cot x \dfrac{dy}{dx} = y(y + 1)$, $-1.3 < x < 1.3$

for which $y = 1$ when $x = \dfrac{\pi}{3}$.

Give your answer in the form $y = f(x)$.

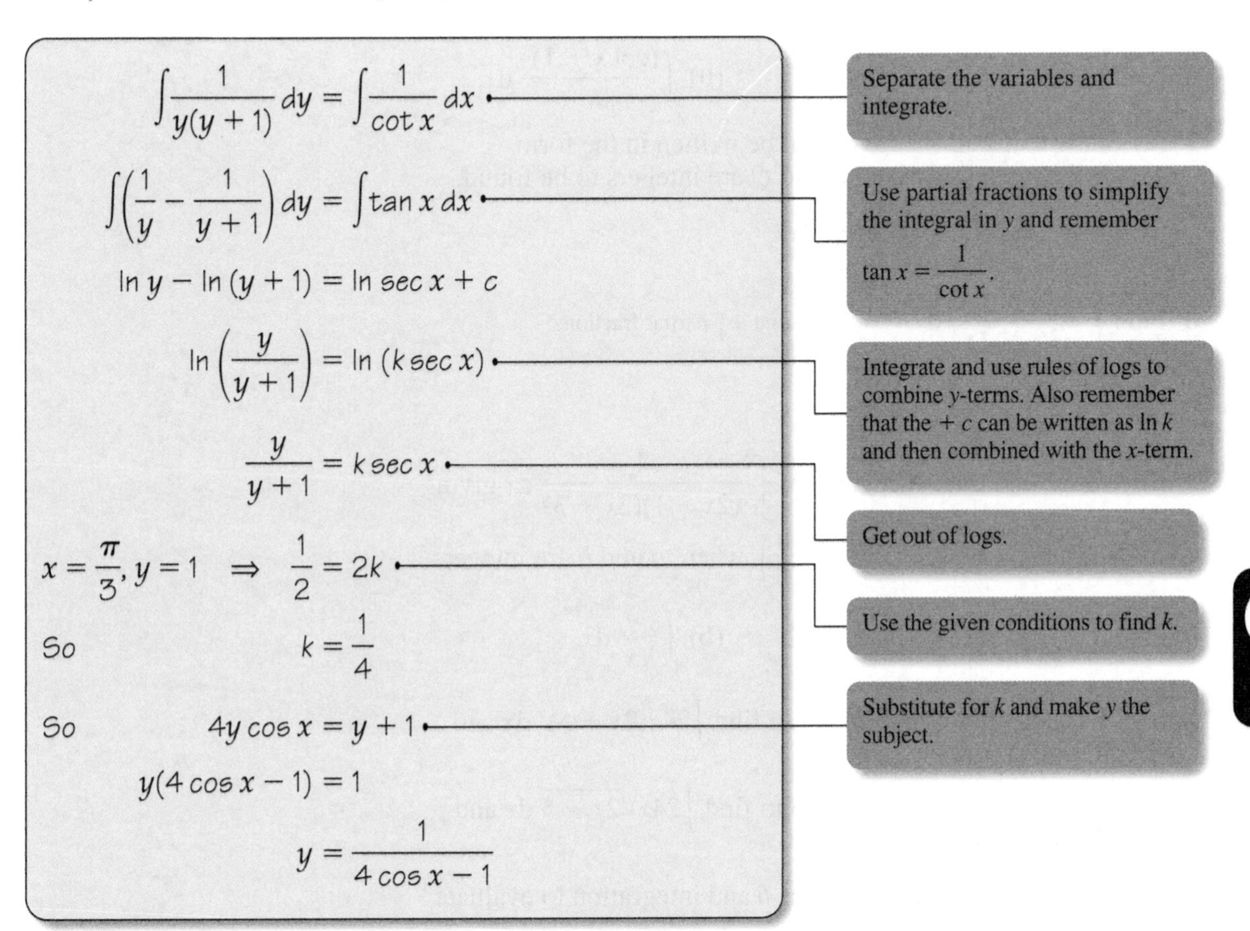

Revision exercise 6

1 Find: **(a)** $\int 2 \cos (3x + 1)\, dx$ **(b)** $\int \dfrac{12}{3x + 5}\, dx$.

2 Find: **(a)** $\int (e^{3x-1} - \sec^2 x)\, dx$ **(b)** $\int \left(\dfrac{6}{3x - 1} + (3x - 1)^2\right) dx$.

3 Find: **(a)** $\int (\cos^2 2x - \sin^2 2x)\, dx$ **(b)** $\int \sin^2 (2x + 1)\, dx$.

4 **(a)** Find $\int \tan\left(2x + \frac{\pi}{6}\right) \mathrm{d}x$.

(b) Hence find the exact value of $\int_0^{\frac{\pi}{12}} \tan\left(2x + \frac{\pi}{6}\right) \mathrm{d}x$.

5 Find: **(a)** $\int \frac{1 - 2x}{5 + x - x^2} \mathrm{d}x$ **(b)** $\int \frac{\sin 3x}{\cos 3x + 2} \mathrm{d}x$.

6 Find: **(a)** $\int x(5 + x^2)^3 \mathrm{d}x$ **(b)** $\int \frac{(\cot x + 1)^3}{\sin^2 x} \mathrm{d}x$.

7 **(a)** Show that $2 \sin 4x \cos 6x$ can be written in the form $\sin Px - \sin Qx$, where P and Q are integers to be found.

(b) Hence find $\int \sin 4x \cos 6x \mathrm{d}x$.

8 Find $\int \frac{2x^2 + x + 1}{x^2(x + 1)} \mathrm{d}x$.

Hint: first use partial fractions.

9 **(a)** Find $\int \frac{4}{(2x - 1)(2x + 3)} \mathrm{d}x$.

(b) Hence find the exact value of $\int_1^3 \frac{4}{(2x - 1)(2x + 3)} \mathrm{d}x$ giving your answer in the form $\ln\left(\frac{a}{b}\right)$, where a and b are integers.

10 Find: **(a)** $\int \sin 2x \, \mathrm{e}^{\cos 2x} \mathrm{d}x$ **(b)** $\int \frac{\mathrm{e}^{\sqrt{x}}}{\sqrt{x}} \mathrm{d}x$.

11 Use the substitution $u = 2x + 5$ to find $\int 24x(2x + 5)^4 \mathrm{d}x$ and simplify your answer.

12 Use the substitution $u^2 = 2x + 5$ to find $\int 24x\sqrt{2x + 5} \, \mathrm{d}x$ and simplify your answer.

13 Use the substitution $u^2 = 1 - \cos\theta$ and integration to evaluate

$$\int_0^{\frac{\pi}{2}} \sin 2\theta \sqrt{1 - \cos\theta} \, \mathrm{d}\theta.$$

14 Use integration by substitution to find $\int_{-3}^{-2} x(4 + x)^3 \mathrm{d}x$.

15 Find $\int x \sin 2x \, \mathrm{d}x$.

16 Find $\int x \operatorname{cosec} x \cot x \, \mathrm{d}x$.

17 Find the exact value of $\int_1^2 x^4 \ln x \, \mathrm{d}x$.

18 **(a)** Find $\int x^2 \sin x \, \mathrm{d}x$. **(b)** Hence find the exact value of $\int_0^{\frac{\pi}{2}} x^2 \sin x \, \mathrm{d}x$.

19 Find the exact value of $\int_0^{\frac{\pi}{4}} \sec^2 x \ln(\sec x) \, \mathrm{d}x$.

20 The region R is bounded by the curve with equation $y = \dfrac{1}{\sqrt{2 + 3x}}$, the x-axis and the lines $x = 0$ and $x = 2$.

(a) Find the area of R, giving your answer in the form $k\sqrt{2}$, where k is a rational number.

The region R is rotated through 2π radians about the x-axis to form a solid shape S.

(b) Find the exact volume of S.

21 The region R is bounded by the curve with equation $y = \tan x$, the x-axis and the line $x = \dfrac{\pi}{3}$.

(a) Find the exact area of R.

The region R is rotated through 2π radians about the x-axis to form a solid shape S.

(b) Find the exact volume of S.

22 The curve C has parametric equations $x = \tan t$, $y = \sin 2t$ and $-\dfrac{\pi}{2} < t < \dfrac{\pi}{2}$.

The region R is bounded by C, the x-axis and the line $x = 1$.

(a) Find the area of R.

(b) Find the volume of the solid formed when R is rotated through 2π radians about the x-axis.

23 **(a)** Complete the following table, giving your answers to 3 decimal places.

x	0	1.5	3	4.5	6	7.5	9
$e^{\sqrt{x}}$	1		5.652			15.466	20.086

(b) Use the trapezium rule and your answers to part **(a)** to estimate the value of $\int_0^9 e^{\sqrt{x}}\,dx$.

24 $I = \int_0^{\pi} \sqrt{1 + \sin x}\,dx$

(a) Use the trapezium rule with 3 strips of equal width to estimate the value of I.

(b) Use the trapezium rule with 4 strips of equal width to estimate the value of I.

The exact value of I is 4.

(c) Comment on your answers to parts **(a)** and **(b)**.

25 **(a)** Find the particular solution of the differential equation
$x^2 \frac{dy}{dx} = \tan y$ where $x = 1$ when $y = \frac{\pi}{2}$.

(b) Find the value of y when $x = 0.5$, giving your answer to 3 significant figures.

26 At time t minutes an ink stain has area A cm^2. When $t = 1, A = 1$ and the rate of increase of A is given by $t^2 \frac{dA}{dt} = A\sqrt{A}$.

(a) Find an expression for A in terms of t.

(b) Show that A never exceeds 4.

27 Use the substitution $x = \tan \theta$ to show that

$$\int_0^1 \frac{1}{(1 + x^2)^2} dx = \frac{\pi}{8} + \frac{1}{4}.$$

Examination style paper

You may use a calculator when answering this paper.
You must show sufficient working to make your methods clear.
Answers without working may gain no credit.

1 Use the substitution $x = 2 + u$ to find the exact value of

$$\int_3^6 \frac{x}{\sqrt{(x-2)}}\,\mathrm{d}x.$$ **(6 marks)**

2 $I = \int_0^1 \mathrm{e}^{x^2}\,\mathrm{d}x$

(a) Given that $y = \mathrm{e}^{x^2}$. Complete the table with the values of y corresponding to $x = 0.2, 0.4, 0.6$ and 0.8.

x	0	0.2	0.4	0.6	0.8	1
y	1					e

(2 marks)

(b) Use the trapezium rule with all the values of y in the completed table to obtain an estimate for the original integral I, giving your answer to 4 significant figures. **(4 marks)**

3 Solve the differential equation

$$\frac{\mathrm{d}y}{\mathrm{d}x} = \frac{3y^2}{(2x+1)^3},$$

given that $y = 2$ when $x = 1$.
Give your answer in the form $y = \mathrm{f}(x)$. **(7 marks)**

4 A curve is described by the equation

$$x^3 + xy - 4x + y^2 - 3 = 0.$$

Find an equation of the normal to the curve at the point (1, 2), giving your answer in the form $ax + by + c = 0$, where a, b and c are integers. **(9 marks)**

5 The line l_1 has vector equation $\mathbf{r} = 3\mathbf{i} + 15\mathbf{j} + a\mathbf{k} + \lambda(7\mathbf{i} + 7\mathbf{k})$ and the line l_2 has vector equation $\mathbf{r} = 5\mathbf{i} - 3\mathbf{j} - \mathbf{k} + \mu(\mathbf{i} + 9\mathbf{j} - 4\mathbf{k})$, where λ and μ are parameters.

Given that the lines l_1 and l_2 intersect:

(a) find the coordinates of their point of intersection, **(4 marks)**

(b) find the value of a. **(3 marks)**

Given that θ is the acute angle between l_1 and l_2,

(c) find the value of $\cos\theta$.
Give your answer as a simplified fraction. **(3 marks)**

6 The function $f(x)$ is defined by $f(x) = (1 - 2x)\sec^2 x$.

(a) Find by differentiation the exact value of $f'(x)$ when $x = \frac{\pi}{4}$. **(5 marks)**

(b) Find an exact value for $\int_0^{\frac{\pi}{4}} f(x)\,dx$. **(6 marks)**

7 Given that $f(x) = \dfrac{9(1+x)}{(1-x)(1+2x)^2}$, $\quad |x| < \frac{1}{2}$,

(a) express $f(x)$ in partial fractions as

$\dfrac{A}{(1-x)} + \dfrac{B}{(1+2x)} + \dfrac{C}{(1+2x)^2}$, where A, B and C are constants to be found. **(4 marks)**

(b) Use the binomial theorem to expand $f(x)$ in ascending powers of x, up to and including the term in x^3, simplifying each term. **(7 marks)**

8

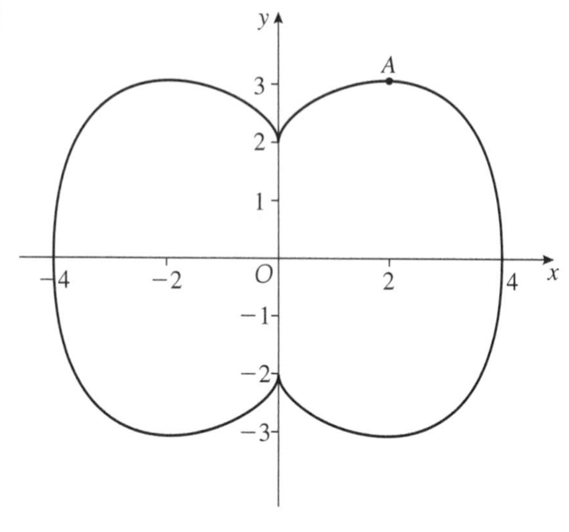

The curve, shown, has parametric equations

$x = 3\sin t - \sin 3t,\ y = 3\cos t - \cos 3t,\ 0 < t \leqslant 2\pi.$

(a) Find an expression for $\dfrac{dy}{dx}$ in terms of t, and show that it may be written as $\cot 2t$. **(4 marks)**

(b) Find the coordinates of the point A where the gradient is zero. **(3 marks)**

(c) Find the total area enclosed by the curve, as a multiple of π. **(8 marks)**

Answers to revision exercises

Revision exercise 1

1 $\frac{3}{2x+1} - \frac{2}{x-3}$ **2** $\frac{3}{x-2} + \frac{4}{(x-2)^2}$

3 $x + 3 + \frac{2}{(x+4)} + \frac{1}{(x-2)}$

4 (a) $\frac{4}{(x+3)} - \frac{1}{(x-2)}$ **(b)** $-\frac{8}{9}$

5 (a) $\frac{1}{(x+2)} - \frac{3}{(2x-1)} + \frac{3}{(x-1)}$

(b) proof

Hint: $\int \frac{A}{(x+2)} \, dx = A \ln (x+2)$

6 (a) $\frac{1}{(x-2)} - \frac{2}{(x-2)^2} + \frac{3}{(2x+3)}$

(b) proof

Hint: $\frac{d}{dx}\left(\frac{1}{x-2}\right) = -\frac{1}{(x-2)^2}$

7 (a) $\frac{4}{(2+x)} + \frac{5}{(1-x)}$ **(b)** $7 + 4x + \frac{11}{2}x^2$ **(c)** $|x| < 1$

8 (a) $\frac{2}{x-3} - \frac{4}{(2-x)} + \frac{1}{(2-x)^2}$ **(b)** proof

9 (a) $A = 2, B = -1, C = -3$ **(b)** $2 + \ln \frac{2}{7}$

Revision exercise 2

1 (a) 16 **(b)** $y = 2x\sqrt{x} - 16$

2 (a) $(-8, 2), (7, -3)$ **(b)** $t = \sqrt{\frac{x+9}{4}}$ **(c)** 9

3 (a) $(0, -3)$ **(b)** $(-3, 0), (3, 12)$ **(c)** $a = 1, b = -4$

4 (a) $\frac{4}{9}$ **(b)** $(0, \frac{27}{2})$

5 (a) $(x - 13)^2 + (y - 5)^2 = 13^2$

(b)

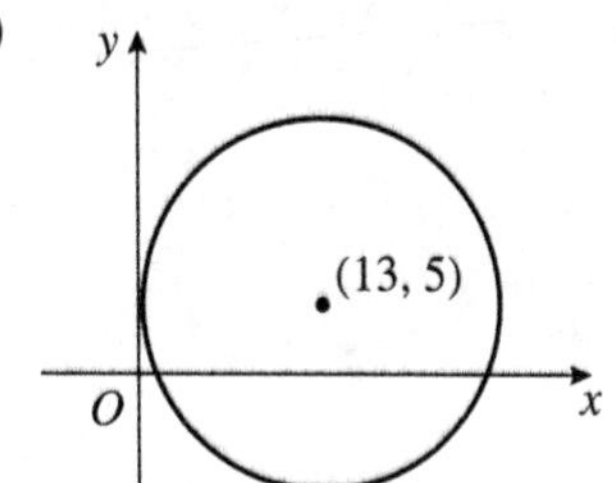

(c) (1, 0), (25, 0)

6 **(a)**

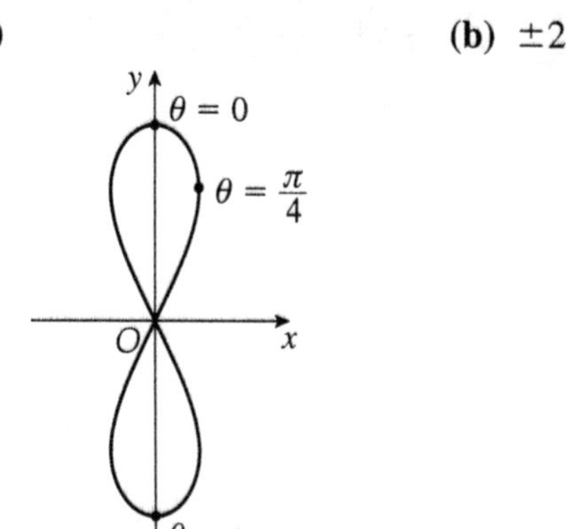

(b) ± 2

7 $A(-\sqrt{3}, 0)$, $B(\sqrt{3}, 0)$

8 **(b)** $f(2) = 0$, $(t + 1)(t - 2)^2$ **(c)** $(-2, -5)$, $(16, 7)$

9 **(a)** $t = \dfrac{1}{y} - 4$ **(b)** $x + 8y - 2 = 0$

10 **(a)** $(-3, 0)$, $(0, 3\sqrt{3})$ **(c)** $\frac{18}{5}\sqrt{3}$

11 **(a)** $(a, 4a)$, $(4a, a)$ **(c)** $4a^2 \ln 4$ **12** **(a)** $2\sqrt{2}$ **(b)** $\frac{64}{3}\sqrt{2}$

13 **(a)** $-2, 0, 2$ **(c)** $17\frac{1}{15}$ **14** **(a)** $(2\pi, 0)$ **(c)** 3π

15 **(a)** 8 **(b)** 32 **(c)** $\frac{9}{64}$

16 **(a)** $-\frac{1}{4}$ **(b)** $y - \dfrac{\sqrt{3}}{2} = 4(x - \sqrt{3})$ **(c)** $x = 1$

Revision exercise 3

1 **(a)** $1 + \frac{1}{2}x - \frac{1}{8}x^2 + \frac{1}{16}x^3 + \ldots$ Valid $|x| < 1$

(b) $1 + 6x + 27x^2 + 108x^3 + \ldots$ Valid $|x| < \frac{1}{3}$

(c) $\dfrac{1}{2} + \dfrac{x}{16} + \dfrac{3}{256}x^2 + \dfrac{5}{2048}x^3 + \ldots$ Valid $|x| < 4$

2 **(a)** $1 - x - \dfrac{x^2}{2} - \dfrac{x^3}{2}$ **(b)** $3 - 2x - \frac{5}{2}x^2 - 2x^3$

3 **(a)** $\dfrac{1}{2} + \dfrac{x}{4} + \dfrac{x^2}{8} + \dfrac{x^3}{16}$ **(b)** $-\frac{1}{2} + \frac{5}{4}x + \frac{9}{8}x^2 + \frac{9}{16}x^3$

4 $a = \pm 2, \mp 32x^3$ **6** $\frac{256}{147}$ **7** $1 + x^2 + \frac{3}{2}x^4 + \frac{5}{2}x^6$

8 **(a)** $a = 6, n = \frac{1}{2}$ **(b)** $\frac{27}{2}$ **(c)** $|x| < \frac{1}{6}$

9 **(a)** $\dfrac{3}{4-x}+\dfrac{2}{1+2x}$ **(b)** $\frac{11}{4}-\frac{61}{16}x+\frac{515}{64}x^2$ **(c)** $|x|<\frac{1}{2}$

10 **(a)** $A=2, B=1, C=-3$ **(b)** $\frac{3}{2}+\frac{29}{4}x+\frac{29}{8}x^2$ **(c)** $|x|<\frac{2}{3}$

11 $\frac{1}{2}+\frac{3}{16}x+\frac{27}{256}x^2+\frac{135}{2048}x^3+\ldots$

Revision exercise 4

1 $(\frac{9}{4},\frac{1}{8})$ **3** $y=1+(\frac{1}{3}\ln 3)x$ **4** $y=-\frac{5}{7}x+\frac{19}{7}$

6 $(1, 1)$ and $(3^{\frac{1}{3}}, -3^{\frac{1}{3}})$ **7** $\dfrac{dy}{dx}=\dfrac{y-2e^{2x}}{2e^{2y}-x}$ **8** -1

10 $\dfrac{dx}{dt}=kx$ **11** $\dfrac{dm}{dt}=-km$ **12** $\dfrac{dT}{dt}=-k(T-22)$

15 $k=\frac{1}{10}$

16 **(a)** $\dfrac{dx}{dt}=-2\sin t+2\cos 2t,\quad \dfrac{dy}{dt}=-\sin t-4\cos 2t$

(b) $\frac{1}{2}$

(c) $y+2x=\dfrac{5\sqrt{2}}{2}$

17 $t=-3$

Revision exercise 5

1 **(a)** $-\mathbf{a}+\mathbf{b}+\mathbf{c}$ **(b)** $-\frac{3}{2}\mathbf{a}-\frac{1}{2}\mathbf{b}+\frac{1}{2}\mathbf{c}$ **2** **(b)** 30

3 14.1 **4** -5 or 8 **5** 34 **6** 24.1°

7 **(a)** $\sqrt{2t^2-12t+42}$ **(b)** 3 **(c)** $2\sqrt{6}$

8 **(a)** $\overrightarrow{BP}=3\mathbf{a}-\mathbf{b},\ \overrightarrow{AQ}=-\mathbf{a}+2\mathbf{b}$

(b) $3\lambda\mathbf{a}+(1-\lambda)\mathbf{b},\ (1-\mu)\mathbf{a}+2\mu\mathbf{b}$

(c) $BX:BP=1:4,\ AX:XQ=2:3$

9 **(a)** $\mathbf{r}=\begin{pmatrix}1\\2\\5\end{pmatrix}+t\begin{pmatrix}-3\\2\\4\end{pmatrix},\ \mathbf{r}=\begin{pmatrix}3\\4\\-1\end{pmatrix}+s\begin{pmatrix}3\\-12\\6\end{pmatrix}$ **(b)** $(4, 0, 1)$

10 **(a)** $(2, 5, 4)$ **(c)** $\frac{3}{2}\sqrt{14}$ **11** **(a)** $-\frac{4}{9}$ **(b)** $\frac{1}{2}\sqrt{65}$ **(d)** $2:1$

12 **(a)** $3\mathbf{i}+6\mathbf{j}+6\mathbf{k}$ **(b)** $-\frac{2}{3}$ **(d)** 2 **(e)** $(2, 4, -5)$

Revision exercise 6

1 **(a)** $\frac{2}{3}\sin(3x+1)+c$ **(b)** $4\ln|3x+5|+c$

2 **(a)** $\frac{1}{3}e^{3x-1}-\tan x+c$ **(b)** $2\ln|3x-1|+\frac{1}{9}(3x-1)^3+c$

3 **(a)** $\frac{1}{4}\sin 4x+c$ **(b)** $\frac{1}{2}x-\frac{1}{8}\sin(4x+2)+c$

4 **(a)** $\dfrac{1}{2}\ln\left|\sec\left(2x+\dfrac{\pi}{6}\right)\right|+c$ **(b)** $\frac{1}{4}\ln 3$

5 **(a)** $\ln|5 + x - x^2| + c$ **(b)** $-\frac{1}{3}\ln|\cos 3x + 2| + c$

6 **(a)** $\frac{1}{8}(5 + x^2)^4 + c$ **(b)** $-\frac{1}{4}(\cot x + 1)^4 + c$

7 **(a)** $P = 10, Q = 2$ **(b)** $-\frac{1}{20}\cos 10x + \frac{1}{4}\cos 2x + c$

8 $2\ln|x + 1| - \dfrac{1}{x} + c$ **9** **(a)** $\dfrac{1}{2}\ln\left|\dfrac{2x - 1}{2x + 3}\right| + c$ **(b)** $\ln\left(\dfrac{5}{3}\right)$

10 **(a)** $-\frac{1}{2}e^{\cos 2x} + c$ **(b)** $2e^{\sqrt{x}} + c$ **11** $(2x + 5)^5(2x - 1) + c$

12 $\frac{8}{5}(2x + 5)^{\frac{3}{2}}(3x - 5) + c$ **13** $\frac{8}{15}$ **14** -8.8

15 $-\frac{1}{2}x\cos 2x + \frac{1}{4}\sin 2x + c$

16 $-x\operatorname{cosec} x - \ln|\operatorname{cosec} x + \cot x| + c$ **17** $\frac{32}{5}\ln 2 - \frac{31}{25}$

18 **(a)** $-x^2\cos x + 2x\sin x + 2\cos x + c$ **(b)** $\pi - 2$

19 $\ln\sqrt{2} + \dfrac{\pi}{4} - 1$ **20** **(a)** $\dfrac{2\sqrt{2}}{3}$ **(b)** $\dfrac{\pi}{3}\ln 4$

21 **(a)** $\ln 2$ **(b)** $\dfrac{\pi}{3}(3\sqrt{3} - \pi)$ **22** **(a)** $\ln 2$ **(b)** $\dfrac{\pi}{2}(\pi - 2)$

23 **(a)** 3.403, 8.342, 11.582 **(b)** 82.48

24 **(a)** 3.908 **(b)** 3.949 **(c)** more strips increases accuracy

25 **(a)** $\ln\sin y = 1 - \dfrac{1}{x}$ **(b)** 0.377

26 **(a)** $A = \dfrac{4t^2}{(1 + t)^2}$ **(b)** since for $t > 0$, $t < (1 + t)$ so $A < 4$.

Examination style paper

1 $8\frac{2}{3}$

2 **(a)** 1.04, 1.17, 1.43, 1.90 **(b)** 1.480

3 $y = \dfrac{12(2x + 1)^2}{9 + 5(2x + 1)^2}$

4 $y - 5x + 3 = 0$

5 **(a)** (7, 15, 7) **(b)** $a = 3$ **(c)** $\cos\theta = \frac{5}{14}$

6 **(a)** -2π **(b)** $1 - \dfrac{\pi}{2} + \ln 2$

7 **(a)** $\dfrac{2}{1 - x} + \dfrac{4}{1 + 2x} + \dfrac{3}{(1 + 2x)^2}$

(b) $9 - 18x + 54x^2 - 126x^3 + \ldots$

8 **(b)** $(\sqrt{2}, 2\sqrt{2})$

(c) 12π

Test yourself answers

Chapter 1

1 $\dfrac{4}{(x+5)} - \dfrac{3}{(x-4)}$ **2** $\dfrac{2}{(x+2)^2} + \dfrac{3}{(x+2)} + \dfrac{1}{(x-4)}$

3 $x - 2 + \dfrac{2}{(x-3)} + \dfrac{2}{(x+1)}$ **4** **(a)** $\dfrac{3}{x} + \dfrac{4}{(x+1)} + \dfrac{2}{(x-1)}$ **(b)** $\dfrac{dy}{dx} = \dfrac{-3}{x^2} - \dfrac{4}{(x+1)^2} - \dfrac{2}{(x-1)^2}$

Chapter 2

1 $y = 4x^2$ **2** $(-8, -2)$ **3** $(x+5)^2 + (y-3)^2 = 1$

4 **(a)**

t	-2	-1	-0	1	-2
$x = t^2 - 1$	-3	-0	-1	0	-3
$y = \frac{1}{2}(t - t^3)$	-3	-0	-0	0	-3

(b) $\frac{4}{15}$

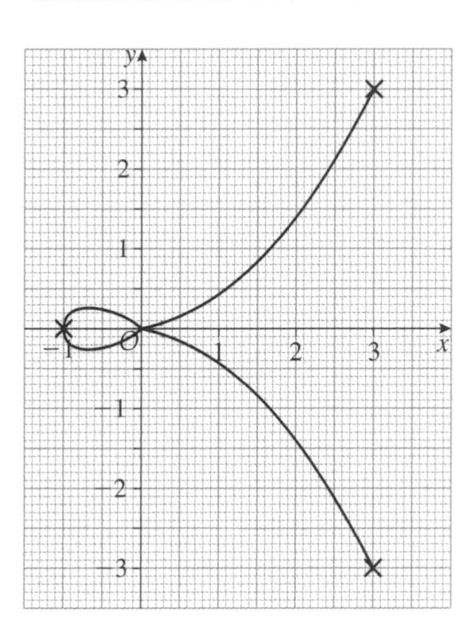

Chapter 3

1 $1 - 6x + 27x^2 - 108x^3$, $|x| < \frac{1}{3}$ **2** **(a)** $3 - \dfrac{x}{6} - \dfrac{x^2}{216} - \dfrac{x^3}{3888}$ **(b)** $9 + \frac{5}{2}x - \frac{13}{72}x^2 - \frac{7}{1296}x^3$, $|x| < 9$

3 **(a)** $b = \pm 24$ **(b)** ± 6

4 **(a)** Proof **(b)** 3.16 **5** **(a)** $A = -4, B = 6$ **(b)** $\frac{5}{3}x - \frac{5}{18}x^2 + \frac{35}{106}x^3$ **(c)** $|x| < 2$

Chapter 4

1 **(a)** $3t$ **(b)** $\dfrac{e^{2t}+1}{e^{2t}-1}$ **(c)** $\dfrac{\cos t}{2\cos 2t}$ **2** **(a)** $y\sqrt{3}+4x=1$ **(b)** $y=x+1$

3 **(a)** $\dfrac{3-x}{3y}$ **(b)** $\dfrac{-4x(x^2+2y^2)}{8x^2y+1}$

4 **(a)** $\ln 2.2^x$ **(b)** $3^x + x\ln 3.3^x$ **(c)** $\frac{1}{2}x^{-\frac{1}{2}}\ln 4.4^{\sqrt{x}}$ **5** **(a)** $16\pi r$ **(b)** 32π **6** $\dfrac{dy}{dx}=\dfrac{1}{9}y^2$

Chapter 5

1 **(a)** $4\frac{1}{2}$ **(b)** -2 **2** $\frac{1}{15}(2\mathbf{i}-14\mathbf{j}+5\mathbf{k})$ **3** **(a)** -2 **(b)** $92.4°$

4 **(a)** $\mathbf{a}-\mathbf{b}+\mathbf{c}$ **(b)** $\frac{1}{2}\mathbf{a}-\frac{1}{2}\mathbf{b}+\mathbf{c}$

5 **(a)** $\mathbf{r}=\begin{pmatrix}2\\7\\-1\end{pmatrix}+t\begin{pmatrix}-3\\4\\2\end{pmatrix}$ **(b)** $c=\frac{1}{2}, d=9$ **6** **(a)** $\frac{7}{25}$ **(b)** $8\sqrt{2}$

Chapter 6

1 $\frac{2}{3}\sin(3x-1)+\frac{1}{3}\ln|3x-1|+c$ **2** $\tan x + 2\ln\sec x + c$ **3** $\frac{1}{2}x^2+\frac{1}{2}\ln|x^2-1|+c$

5 $\frac{1}{3}xe^{3x}-\frac{1}{9}e^{3x}+c$ **6** $y=\dfrac{6\sin x-1}{6\sin x+1}$